Earl of Dundonald

**A Treatise**

Shewing the intimate connection that subsists between agriculture and chemistry,

addressed to the cultivators of the soil, to the proprietors of fens and mosses, in

Great Britain and Ireland; and to the proprietors of West India estates

Earl of Dundonald

**A Treatise**
*Shewing the intimate connection that subsists between agriculture and chemistry, addressed to the cultivators of the soil, to the proprietors of fens and mosses, in Great Britain and Ireland; and to the proprietors of West India estates*

ISBN/EAN: 9783337325138

Printed in Europe, USA, Canada, Australia, Japan

Cover: Foto ©berggeist007 / pixelio.de

More available books at **www.hansebooks.com**

A

# ʼTREATISE,

SHEWING THE

## INTIMATE CONNECTION

THAT SUBSISTS BETWEEN

# AGRICULTURE AND CHEMISTRY.

ADDRESSED TO THE

## CULTIVATORS OF THE SOIL,

TO THE

## PROPRIETORS OF FENS AND MOSSES,

IN

## *GREAT BRITAIN AND IRELAND;*

AND TO THE

## *PROPRIETORS OF WEST INDIA ESTATES.*

BY

# THE EARL OF DUNDONALD.

---

" Let us cultivate the Ground. that the Poor, as well as the Rich, may be filled, and Happiness and Peace be established throughout our Borders."

---

*LONDON*

PRINTED FOR J MURRAY AND S. HIGHLEY, (SUCCESSORS TO THE LATE MR. MURRAY), NO. 32, FLEET-STREET.

---

1795.
(Drawback.)

# HEADS, OR TABLE OF CONTENTS.

ERRATA.

## ERRATA.

Page 7, 3d line from the top, *for* angusti *read* angusta

—— 9, 2d line from the bottom, *for* qui *read* quæ

—— 35, 6th line from the bottom, prefix power of *to the word* solution

—— 40, 3d line from the bottom, **read** thereto *instead of* there to

—— 43, 6th line from the top, *read* elective *instead of* electric

—— 54, 13th line from the top, *read* colour *instead of* colou

—— 75, 11th line from the top, *read* elective *instead of* electrive, an error continued throughout the Work.

—— 103, 10th line from the top, *for* œconomies *read* œconomics

—— 115, 8th line from the bottom, *for* produced *read* occasioned

# INTRODUCTION.

THERE are at present a variety of obstacles to the advancement of Agriculture in these kingdoms, or to the production of the greatest quantity of food from the soil. Amongst this variety there are those of a nature not to be removed but by the arm of Government; whilst there are others which only require due exertions on the part of individuals.

The slow progress which Agriculture has hitherto made as a science, is to be ascribed to a want of education on the part of the cultivators of the soil, and the want of knowledge, in such Authors as have written on Agriculture, of the intimate connection that subsists between this science and that of Chemistry. Indeed there is no operation or process, not merely mechanical, that does not de-

A pend

pend on Chemistry, which is defined to be a knowledge of the properties of bodies, and of the effects resulting from their different combinations.

In the following pages an attempt will be made to explain, on established principles, the processes that accompany the cultivation and amelioration of the soil. This discussion will come forward with peculiar advantage at a time when provisions bear so high a price, and when individuals, awakening from the golden dreams of manufactures and of commerce, begin to see, and experimentally to feel, that the prosperity of a nation cannot be permanent, nor its inhabitants quiet and contented, in their respective situations, where Agriculture is neglected, and an unwise preference given to manufactures and to commerce ; occupations that produce very different effects on the bodies and minds of men, from those that are attendant on the sober and healthful employment of husbandry.

By the adoption of a new line of investigation, exemplified in the following Treatise, light has not only been

been thrown on the action and effects of the manures at present employed, but the uses of other substances, and methods of combining and preparing them, have been discovered ; from which there are just grounds to believe a valuable addition will accrue to the present stock of Agricultural Knowledge.

It will appear that the saline, and other substances capable of being applied to promote vegetation, are very numerous ; by far the greater part of them have escaped the notice of those who have made Agriculture their study, nor have any attempts even been made to explain on chemical principles the operation of the manures and substances now in use. Had such researches been prosecuted to effect, they would have led to the discovery and application of other substances capable of being employed, with equal, or perhaps superior advantages.

The promoting of Agriculture is not solely to be considered as creating a more plentiful supply of food, but it is to be regarded as morally and politically conducing to the true happiness of man, by giving to him the occu-

pation

pation allotted to his first parents; whence flow health, social order, and obedience to lawful authority; consequences very different from those that are produced by the over-driven system of manufacturing, in which the industrious workman is often subjected to great inconveniencies, not only by the fluctuation in the demand for the articles he manufactures, but likewise by a total suspension of trade by war or other causes. When evils, like these, which he has no power to avert, press hard upon him, he is frequently rendered desirous of assisting to bring about any political change, whereby he is tempted to believe that his situation may be rendered more comfortable; hence he becomes unquiet, and to society a less valuable member than the husbandman, whose occupation does not expose him to such distress, nor to the like temptations.

Whilst the benevolent must feel for the hardships to which, at times, manufacturers are liable, still they cannot but recollect the restless spirit so frequently manifested by persons of this description, even in the moment of the greatest prosperity; especially in towns and cities,

cities, where, for the interest and convenience of the masters, and not of the workmen, they are collected in great numbers.

To this assemblage may be ascribed the dissemination of pernicious doctrines, by a few profligate persons, who are to be found in all societies, and who have it in their power to corrupt the good principles of the many. The well affected thus become the tools of the seditious and designing.

To such political evils there is but one remedy :

That a preference to all other pursuits be given to Agriculture.

That the establishment of such branches of manufacture, as it may be wise to encourage, be promoted only in scattered villages, resembling the townships in America. By this plan the diseases of the body and the mind would be rendered less contagious ; each individual might, at a moderate rent, be supplied not only with a sufficient extent of ground, to enable him to keep a cow, and supply

his

his family with milk, an article indispensably necessary to the rearing of healthy children ; but also what more might be requisite for raising potatoes, and other vegetables, cultivated at his leisure by the spade, affording an agreeable and healthful change to his confined and sedentary occupation.

The Legislature of these kingdoms---all good men---and all well disposed subjects, are earnestly called upon to unite in promoting the more complete cultivation of the soil; being the only system by which the comforts and the happiness of the people can be advanced, and the future existence of this country, as a kingdom, be effectually secured against foreign foes and domestic incendiaries.

PREFATORY

# PREFATORY ADDRESS.

MANY experiments, much labour and thought, a desire to be usefully employed, with the leisure afforded by a retired life, proceeding from the *res angusta domi*, have enabled the Author to present the following Treatise to the Public.

The importance of such a Work, and the peculiar fitness of it at this juncture, have induced him of late to dedicate the whole of his time to its compilation and arrangement; accomplished with much personal inconvenience, and to the apparent neglect of some important duties and family concerns, which otherwise would have claimed his prior regard.

The

The Author flatters himself, that his labours will be found to open a field of experiment, of chemical reasoning, and of the practically useful, applicable to Agriculture, of which that science has hitherto been thought incapable.

In the prosecution of his researches to so desirable an end, he acknowledges the assistance he has derived from the labours and valuable discoveries of a PRIESTLEY and a CAVENDISH, under the heads of Air, or Gasses, Composition, and Decomposition of Water ; so true is the observation, that there is no art or discovery that is not the parent or sister of some other.

Gratitude requires that he should likewise acknowledge the obligations he is under to his friend Mr. VANCOUVER, for the time he has kindly dedicated, and the relief he has afforded the Author in the fatigue of this publication---a work wherein accuracy of chemical expression---a variety of important matter---attended with the difficulty of distinct arrangement, never completely to be attained in any work, where the subjects to be

treated

treated of are so connected and interwoven the one with the other, has necessarily required much reflection. Great care has been taken, in the arrangement adopted, to lead the mind of the reader from the discussion of such substances as are simple, to those that are of a compound nature; the practically useful observations unfolding themselves as the work advances.

Contrary to the system of modern oratory and book-making, perspicuity, and its constant attendant, brevity, have been uniformly adhered to, under the conviction, that clear ideas are best expressed by the fewest words.

" *Non asseveravi quæ vastitas hujus scientiæ contineret cuncta me dicturum sed quædam : nam illud in unius hominis prudentiam cadere non poterat ; neque enim est ulla disciplina aut ars, qui singulari consummata sit ingenio.*"

COLUMELLA, lib. v. f. 166.

B

# A

# TREATISE,

SHEWING THE

## INTIMATE CONNECTION

THAT SUBSISTS BETWEEN

# AGRICULTURE AND CHEMISTRY.

---

## EARTH.

By the word earth, in its common acceptation, is either to be understood the habitable globe—the solid or dry part of it, in opposition to the aqueous, or the surface stratum, soil, or mould, in opposition to the more solid under strata ; whilst the word earth, taken chemically, signifies a dry, uncompounded, simple substance, incapable of being volatilized, or acted upon by fire.

There

There are different kinds of such simple, or, as they are frequently called, primary earths; those which generally occur, and principally regard the object of this Treatise, are, calcareous, or chalky; argillaceous, or clayey; siliceous, or sandy; and magnesian. It is also necessary to mention, that the earth of iron is likewise contained in most soils, in great abundance; existing therein, in various states.

––––––––––

## CALCAREOUS MATTER

Constitutes not only the surface, or soil, but likewise the under stratum of many countries, to a very great depth. Under this general name of calcareous matter is included chalk, marble, limestone, coral, shells, &c. The three first mentioned are frequently mixed with iron, and with different proportions of the simple earths; but are considered as calcareous, when the proportion of that earth predominates. It is capable of absorbing, and

of

of retaining moisture, though in a considerably less degree than clay.  By the action of fire it becomes lime,
and returns again to the state of chalk, or calcareous
matter, by exposure to air.

---

## ARGILLACEOUS MATTER

FORMS not only a large portion of the surface soil of
most countries, but is also found in the mineral strata, to
an immense depth.  Argillaceous matter, or clay, is no
where found pure, is more or less adulterated with the
different earths, and with different materials; such as
mineral, vegetable, and animal substances.

The purest clay contains upwards of sixty per cent. of
siliceous matter, or sand.

Clay is the earth most retentive of moisture, by
which it becomes ductile and tenacious; and loses these
properties by the action of fire.

SILI˙

## SILICEOUS MATTER.

GREAT tracts of the surface of the earth are of this description, and very large masses of the under stratum consist of the like substance : the former in the state of loose sand, and the latter in an indurated or solid state, called sand-stone or free-stone. It is, of all the earths, the least retentive of moisture.

---

## MAGNESIAN EARTH.

THIS is no where found in such quantities as to form a soil of itself; it is contained in different proportions, in many soils, and constitutes a component part of stea-tites, or soap rock. It is retentive of moisture to a certain degree.

EARTH

## EARTH OF IRON

Exists in ground in a metallic state, in the state of an earth, and in the state of a mineral salt. It is the only metallic earth thought necessary to define, in this Treatise on Agriculture. An attempt will hereafter be made to explain the manner in which it promotes vegetation.

---

## AIR.

By air is generally understood, the medium in which terrestrial animals move and breathe. It is possessed of weight or gravity, is capable of compression: without it neither animals could live, nor could fire be maintained, or heat generated.

Part only of atmospheric air serves for the support of animal life and combustion, called vital air, pure air, or oxygen.

oxygen. The other part, c lled phlogisticated air, or azot, is injurious to combustion, although necessary in a certain proportion, to modify the too powerful action of pure air, in the respiration of animals. Vital air, by combining with ignited inflammable substances, produces certain gasses and new compounds.

Atmospheric air contains likewise a proportion of fixable air, or carbonic acid gas, so called by the French chemists, from carbonaceous matter, or charcoal, forming one of its constituent parts.

Atmospheric air may also contain other gasses, or airs. By far the greater part of these gasses were discovered by Dr. PRIESTLEY.

By the term of gas is to be understood, a permanently elastic, invisible fluid: Of these there are some which may be considered as simple, and others as compounded.

The gasses are, empyreal air, vital air—or—oxygen gas;
Fixable air—or—carbonic acid gas;
Light inflammable air—or—hydrogenous gas;

Dense

Dense inflammable air—or—carbonated hydrogenous gas;

Phlogisticated air—or—azotic gas;

Vitriolic acid air—or—sulphureous gas;

Sulphurated inflammable air—or—sulphurated hydrogenous gas;

Nitrous air—or—nitrous gas;

Muriatic air—or—muriatic gas, &c. &c.

Atmospheric air is equally necessary to the vegetation and growth of plants, as to the life of animals. By a most beautiful arrangement in the œconomy of nature, the different processes of animal and vegetable respiration are made mutually to assist each other. The particular gas or air thrown off by the respiration of the one, contributes to the existence of the other.

## WATER

Is the fluid in which fishes live, and forms part of the support of plants, and of terrestrial animals. It was formerly considered to be a simple uncompounded substance; but it has been discovered by Mr. CAVENDISH to be a compound body, consisting of the basis of inflammable air and vital air. It is formed naturally, and artificially, by uniting the basis of inflammable air with vital air, by means of the electric fluid.

Water is capable, under certain circumstances, of containing, or uniting with, the bases of different gasses; it contains, for the most part, fixable air; to which its power of holding in solution calcareous matter, contained in most water, is to be ascribed.

Water, by putrefaction, and by the processes of vegetation and animalization, is capable of being decomposed, or having its component parts separated.

HEAT.

## HEAT.

By heat is to be understood the sensation arising from the solar rays, or from the combustion of inflammable bodies. Combustion cannot take place without the presence of air, and it is only the vital part of air that promotes this process. A certain degree of heat is requisite for animal and vegetable life. It is generated in animals, in consequence of respiration, by the absorption of the vital part of atmospheric air, and combination thereof with the animal system; here called the process of animal oxygenation: and as animals have thus the power of generating heat by respiration, they can live in a much greater degree of cold than vegetables.

To the presence of heat is to be ascribed the fluidity of water; to the absence of it, its solidity, or the formation of ice.

The principle of heat is contained in bodies without sensibly manifesting its presence; it is then called latent

C 2                                          heat.

heat. This is capable, by certain processes, of being disengaged from those bodies, and of becoming manifest to the senses, or of passing and entering into a state of combination with other bodies.

The frequent changes in the degree of heat and cold in the atmosphere, are to be ascribed more to the alternate disengagement and fixation of heat, by chemical combination, than to the effects of the solar rays.

## SALINE SUBSTANCES

ARE compounds, consisting of different combinations and proportions of earth, gasses, water, and the principle of heat; to which must be added (although not yet treated of) the results of more compounded substances, such as animal and vegetable matter.

Saline substances consist of acids, alkalis, and neutral salts, resulting from the combination of acids with earths and alkalis. These are stated hereafter in separate tables, wherein their uses and properties, as applicable to agriculture, are more fully explained.

VEGETA-

## VEGETABLES

ARE organized bodies, capable of growth and increase, and of propagating their own kinds. Vegetables are nourished, supported, and formed by air, water, earth, heat, light, and certain saline substances; and, in a particular manner, by their own exuviæ, or remains, when reduced to a state fit to answer that purpose.

Vegetables consist of mucilaginous matter, resinous matter, matter analogous to that of animals, and some of a proportion of oil. All these matters serve different and important purposes in the œconomy of plants. The resinous and animalized matters form the outward surface of vegetables, which prevents their being acted upon by rain or moisture.

The mucilage gives pliability; is the principal and first prepared juice of plants, of which, by a further degree

of

of maturation, resin and oil are formed. This oil contributes also to the pliability of vegetables.

By a due mixture of mucilage with these other substances, vegetables are to a certain degree rendered capable of solution in water, a property not possessed either
by resin, animal matter, or oil, in a separate state.

Thus it appears, that although living vegetables are
effectually protected, by their outward covering of resin
and animalized matter, from the action of humidity;
still, from the whole of their constituent parts, there
arises a certain degree of solubility, which afterwards contributes to the food both of animals and vegetables.

Beside these, vegetables contain earthy matters, formerly held in solution in the newly taken in juices of
the growing vegetable.

ANIMA-

## ANIMALIZED MATTER CONTAINED IN VEGETABLES.

THE greatest proportion of the animalized matter found in vegetables, is contained in grain. Grain consists partly of mucilage, or starch, and partly of this substance, called by the French chemists vegeto-animal. These two substances constitute what is termed meal or flour. They are capable of separation, therefore they exist in grain in a state of mechanical mixture, not of chemical union. This union is to be accomplished by the process of germination or malting: the result of which is sacharine matter or sugar. This sacharine matter, by fermentation, is further attenuated and resolved into vinous spirit, and ultimately by exposure to air, and by the absorption of vital air, into the acetous acid, or vinegar. By distillation of the vinous spirit, ardent spirit is obtained. The quantity of sacharine matter, of ardent spirit, and of vinegar capable of being procured from grain, depends upon its containing a due proportion of starch and animalized matter: neither of these substances, taken singly, can be made to yield sacharine matter.

Different

Different kinds of grain contain these two substances in different proportions; and the same sort of grain contains them in different quantities, according to the climate, season, and soil.

Good and well raised bread depends on flour containing a due admixture of these two substances. Hence, by mixing the flour of different sorts of wheat, better bread may at times be produced than from one sort only. Extensive benefits may also accrue to the processes of brewing, distilling, and making of vinegar, by a mixture of the different sorts of grain: to that of distilling, a further advantage would be derived by a mixture of different roots with the grain; such as potatoes, parsnips, carrots, &c. &c. if prepared in a proper manner.

Good wheat generally contains two fifths of animalized matter, and three fifths of starch. The œconomical management of the manufacture of this article is capable of considerable improvement.

VEGE-

## VEGETABLES ANALYSED BY FIRE.

VEGETABLES, by the application of heat, and by distillation in close vessels, are resolved into different gasses, liquid matters, and into insoluble matter.

The gasses consist of inflammable and fixable air: the liquids, of water, vegetable acids, and oil: the insoluble, of charcoal.

By combustion in the open air, charcoal is resolved into fixable air, soluble, and nearly insoluble compounds.

The soluble are, alkaline and neutral salts.

The insoluble, for the most part, consist of lime combined with the phosphoric acid, called phosphat of lime.

By a further, and more intense application of heat, all the constituent parts of vegetables, excepting the

D                                          earthy,

earthy, may be changed into the gassious state, and re--solved into their simple principles.

By combustion in the open air, grain, carrots, potatoes, &c. &c. yield much more alkaline salts, than straw, hay, or wood.

Neither starch, nor the animalized matter of grain, yield fixed alkaline salts, when burned separately; hence it appears, that the union of these substances is equally requisite for the formation of fixed alkaline salts as for sacharine matter.

This analysis of vegetables by fire, is far from shewing the true state of combination, in which the simple principles of these substances existed in vegetables. The acid phlegm, the empyreumatic oil, the vegetable alkali, or potash, vitriolated tartar, or other neutral salts, are compound matters, or new combinations, produced by the action of heat, and decomposition of water, and are substances not only very different from those juices which the plant originally imbibed, but likewise very different from those matters found in the vegetables, after

having

having undergone, by the vegetative process, their full degree of concoction and maturation: whence it is obvious, that neither the precise food of vegetables, nor their component parts, can be ascertained by any analysis of them by fire; at least it is not practicable with the several substances, resulting from such process, to recompose a juice, or fluid, similar to that by which the vegetable had originally been nourished.

It has been stated, that by an intense degree of heat all the component parts of vegetables, excepting the earthy, may be resolved into permanently elastic fluids, or gasses; and into the compound substance called water.

By vegetables being thus reduced to their simple or elementary principles, they are found to be composed of gasses, with a small proportion of calcareous matter. Although this discovery may appear of small moment to the practical farmer, yet it is well deserving of his attention and notice, as it throws great light on the nature and food of vegetables, and proves that a large proportion of vegetables consist of the aery form, fluids or gasses.

ANALYSIS OF VEGETABLES BY PUTREFACTION.

THE resolution or separation of the component parts of vegetables by putrefaction, must evidently appear far better adapted to answer the inquiries of the Chemical Agriculturist, than any analysis of them by fire.

This process can take place only when attended with air, moisture, and a due degree of heat. Water is decomposed—vital air absorbed—heat disengaged, and new combinations formed, such as

Gasses—with soluble and insoluble matters.

The gasses are—Inflammable and azotic or phlogisticated air, forming volatile alkali and fixable air.

The soluble saline matters consist of phosphoric, soreline, or vegetable acid, combined with vegetable, mineral, or volatile alkali.

The

The insoluble matters consist of phosphoric, soreline, or vegetable acid, combined with lime, or calcareous matter.

These last mentioned substances may likewise be produced from vegetables which have not undergone the putrefactive process, by the tendency which pure air, vital air, or oxygen, has to combine with such like, and all other inflammable substances; a process here called

## OXYGENATION.

By the combination of pure air with inflammable substances, particular acids are formed, with the peculiar bases of those acids contained in inflammable substances. The acids, as they are formed, combine either with the calcareous matter of the vegetables, or with other calcareous matter in the soil, forming salts, which for the most part are very insoluble.

The process of putrefaction is always accompanied by that of oxygenation : but oxygenation may be, and is to a great extent, independent of putrefaction.

To

To this process of oxygenation, the continuance of vegetable matter on the surface of the earth is principally to be ascribed; as in the case of peat mosses, fens, and morasses, as well as in most soils, but more especially in such as have long been under cultivation. The indestructible state of vegetable matters, under these circumstances, and their constant accretion, may be referred to the insoluble compounds, produced by the action of pure air on these inflammable substances.

The insolubility, to a certain degree, of this system, adopted by nature, is undoubtedly to be preferred to one more completely soluble; for it is evident, that if putrefaction, or oxygenation, had possessed the power of rendering all the vegetable matter, by a speedy process, soluble in water, two pernicious consequences must have followed.: The rains would have washed down such extracts, and soluble matters, as fast as formed, into the rivers and springs, contaminating the waters, and rendering them unfit for the existence of fishes, or for the use of terrestrial animals. The sea, in process of time, would thereby receive all the vegetable and animal produce of the dry land, and the earth would ultimately become

come barren, consisting alone of the simple earths, without any admixture of vegetable matter; consequently there could be no accumulation of this substance on the surface, as is the case to an immense degree at present. As such there cannot be a doubt, but that the present incomplete process of putrefaction, oxygenation, or solution of organic bodies, has been established by the Great Creator of all things for wise and benevolent purposes; especially when it shall be understood, that the apparent imperfections of this (to a certain degree) insoluble system are, as they respect agriculture and vegetation, to be remedied, when necessary, by the ingenuity and industry of man.

A frequent exposure of fresh surfaces to the action of the air, as in the case of fallowing, will, by promoting oxygenation, increase the insolubility of vegetable matters contained in the soil.

INERT

## INERT VEGETABLE MATTER OR PEAT.

INERT vegetable matter, or peat, is, for the most part, formed of the remains of aquatic vegetables, or of those vegetables which generally grow in humid or moist situations. Their nourishment and growth are promoted by atmospheric air, by the decomposition of water, and by the calcareous matter held in solution, and contained in most water. These substances alone, according to the analyses of vegetables already given, are sufficient to account for the growth of such aquatic vegetables, and the accumulation on the surface of the earth of such matter forming peat mosses and fens.

Trees of a considerable size have been frequently found at the bottom of peat mosses, with the appearance of having been cut down, or in part acted on by fire. Hence it may be inferred, that the peat moss itself did not give birth to, or support the growth of such trees; but on the contrary, that by the destruction of forests, in consequence of natural causes, fire or war, the

the trees had been thrown down, and causing a stoppage of the waters in their passage to the sea, the growth and decay of the aquatic vegetables already described, had formed those extensive peat mosses and fens, which, in their natural state, are of all soils the most unproductive, but which are the most fertile when improved.

Peat is very retentive of moisture, retaining it in a manner similar to that of a spunge. At no time, therefore, in this humid and northern climate, can such soils be divested of their superabundant proportion of moisture with which they charge themselves in the autumn, spring, and winter, as well as during the periodical rains in summer. The sun's rays, or drying winds, during the summer season, are exerted in conveying away, by evaporation, this surplus moisture; and as heat is known to be abstracted from bodies, and cold generated by evaporation, hence effects will arise injurious not only to climate, but likewise to vegetation in general; but more especially so to such plants as require a greater degree of heat and nourishment, than soils of the above description will admit of. There can be little doubt, that these injurious effects on vegetation will extend themselves

E

even

even to the drier lands in the vicinity of such fens or mosses.

The draining, reclaiming, and cultivating, lands so circumstanced, must appear not only important, from the great value of such lands when reclaimed, but likewise from the effects that such drainage would have on the climate, temperature, and vegetation of the adjacent country.

Peat is an inflammable substance; consequently capable of combining with pure air, or oxygen, and of becoming oxygenated; a process already explained in the preceding part of this Treatise. The surface of peat mosses, or what is most exposed to the action of air, is capable of becoming more oxygenated than the under stratum.

The oxygenation of peat, and indeed the combination of pure air or oxygen with inflammable substances, renders such substances less inflammable, a process analagous to that of combustion: in both cases saline compounds are formed, which are uninflammable.

It

It is upon this principle that inflammable bodies when exposed to air lose their combustibility, it being evident, that such matters as had saturated themselves with a full proportion of vital air, or oxygen, are incapable afterwards of combining with a greater quantity, and consequently must be uninflammable. On this account, the upper stratum of peat mosses is generally thrown aside when peats are dug for fuel.

The longer peat moss is kept exposed to air, the less soluble it becomes, and ultimately imparts no colour to water; whilst peat newly formed, or in a less degree oxygenated, imparts a colour to water which will be found to contain the extractive saline matters of fresh or less decayed vegetables.

It is owing to this solution in water that no alkaline salt is procured from the ashes of peat, decayed vegetables, or water soaked wood.

In peat mosses there are frequently springs of mineral water, which contain in solution saline and ferruginous matters. Hence the ashes of peat, besides the earthy

E 2 matter,

matter, (consisting for the most part of phosphat of lime)
contain likewise phospaht of iron, gypsum, Epsom salt,
and green vitriol; and these in different proportions, according to the nature of the peat, and circumstances under
which it had been formed. Hence also the ashes of
different kinds of peat will have different effects when
used as manures, or top-dressings, to ground.

The decayed remains of vegetables, called in this Treatise inert vegetable matter, abundantly contained in many
soils, especially those which have been much manured,
and long under cultivation, are in all respects similar
to peat, and capable, like peat, of different degrees of
oxygenation and insolubility; a process promoted by
fallowing, or the exposure of fresh surfaces to the action
of air.

A method of rendering these inert vegetable matters conducive to vegetation, will be given in the sequel
of this work, when the application of saline matters to
different soils is discussed.

FOSSILE

## FOSSILE COAL

Is an inflammable substance, formed of the remains of antediluvian vegetables, animal juices, and mineral or metallic substances, combined or mixed with earthy matters. Like peat it loses its inflammability by exposure to air, and becomes oxygenated. Saline compounds are thence formed; they consist of green vitriol, Epsom salt, phosphat of lime, phosphat of iron, together with earth of iron, and a proportion of the uncombined simple earths.

Oxygenated fossile coal is likewise capable of solution by saline substances, and of producing the same good effects in promoting vegetation as oxygenated peat, when treated in a similar manner.

Such coal as is most applicable for this purpose is found at the crop or outburst of most seams, particularly those which are of a soft tender nature, and easily acted upon by the joint influence of air and water. It

will

will require many years before coal, (wrought at a considerable depth, and brought to the surface) will, by exposure to air, become oxygenated, and attain what coal at
a less depth had acquired by the action of air and water
during many ages.

When coal is in a state capable of being rendered soluble, it is soft and friable; when rubbed between the
fingers it appears like soot; when thrown into the fire
it does not burn with any flame; and whilst consuming, emits a smell more like unto that from the combustion of peat than coal.

Most fossile coal contains the bases requisite for forming, with a combination of pure air or oxygen, the following acids, viz. the vitriolic or sulphuric, muriatic,
phosphoric, &c.

The acid first formed is the sulphuric. As this is generated, it combines either with the earth of iron or the
earth of magnesia, forming green vitriol and Epsom salt.
These salts being very soluble, are readily washed away
from the coal by rain or moisture; after which, by a
farther

farther supply of oxygen with their peculiar bases, the phosphoric and other acids are formed. These acids, as they are generated, combine with the calcareous matter of the coal, now by full decomposition rendered capable of being acted upon, and form salts that are nearly insoluble.

Oxygenated coal, when it is not found in that state, may be procured by exposing the small or refuse coal of a colliery alternately to air and moisture. This process may be much accelerated by previously grinding the coal to a fine powder.

---

## CHARCOAL, OR THE COKE OF COAL,

Is a substance in an intermediate state between that of vegetables, wood, or fossile coal; and that of their reduction to ashes. The making of charcoal is best performed in close vessels, or by a smothered method of combustion.

By

By the combustion of charcoal in the open air, and in consequence of the combination of vital air, fixable air is produced.

Carbonaceous matter or charcoal is likewise resolved into fixable and inflammable air, by the alternate application of moisture and heat to charcoal, in close vessels, in which case water is decompounded.

There is reason to believe that the solar rays are no where on the surface of the earth sufficiently powerful to form charcoal or coke from vegetable matters or fossile coal.

Vegetable substances contain the carbonaceous principle, or what by heat may become charcoal or coke, but are not prior there to charcoal.

Charcoal, by different processes, may be made to afford the carbonaceous principle to plants.

## SULPHUREOUS SCHYST.

THIS substance generally accompanies the strata of coal and limestone. For the most part it consists of clay and pyrites, *i. e.* iron mineralised by sulphur, with a proportion of calcareous matter and magnesia, the whole tinged with a black, blue, or grey colour, by mineral tar or oil. By exposure to air vitriolic acid is formed; this, by combining with lime, magnesia, earth of iron, or clay, forms gypsum, Epsom salt, green vitriol, or allum.

Sulphureous schyst acts powerfully as a manure to soils containing much phosphat of lime, or calcareous matter. It is found, in the greatest abundance, in the East part of the North Riding of Yorkshire, and is there called allum rock.

F                                                    LIME.

## LIME.

Is produced from chalk, marble, limestone, coral, or shells, when submitted to a sufficient degree of heat to disengage the fixable air, or carbonic acid, with which, in a natural state, these substances are always united.

Water has a great tendency to combine with lime; heat is disengaged, and there is reason to believe that the water is decomposed in the process. Newly made lime, from its power of destroying, or as it were burning vegetable and animal bodies, is termed caustic. When applied to organic bodies, containing moisture, it rapidly destroys their adhesion, or continuity of parts, and disengages from them inflammable air, and azotic or phlogisticated air, forming volatile alkali. The residuum will be found to consist of charcoal, and of a combination of lime with the phosphoric and other acids, forming saline matters, which are nearly insoluble. The above effects are produced by the application of lime to peat, or to soils containing

taining much vegetable matter; part of which is dissipated in a gassious state, and part combines with the lime, forming insoluble compounds, which cannot promote vegetation, until brought into action by other saline substances, either on the principle of superior affinity, or on that of the double electric attractions, as will be explained in the sequel of this work.

## CHALK, OR UNCALCINED CALCAREOUS MATTER.

CHALK has not the same power as lime in destroying the texture of organic bodies, because it is saturated with fixable air: it has, however, an action on these substances, or more properly speaking, these substances have an action on chalk, so soon as by oxygenation their respective acids are generated; in which case they will combine with the chalk, and form the nearly insoluble saline compounds, already described. By the action of

these

these acids on chalk fixable air is disengaged. This gas is absorbed by the roots of vegetables, decomposed in the process of vegetation, affording the carbonaceous principle to vegetables; whilst vital air, the other constituent part of fixable air, is thrown off by the respiratory power of vegetables.

The different effects of lime and of chalk on vegetable and animal substances, shew, that when either of those materials is applied with a view to promote the fertility of the ground, they should be used according to circumstances, and the nature of the soil.

PAR

## PARTICULAR EFFECTS RESULTING FROM THE APPLI-
## CATION OF LIME AND CHALK TO GROUND.

LIME is known to have a tendency to sink below the upper surface, and to form itself into a regular stratum between the fertile and the unfertile mould. After breaking up pasture ground that formerly had been limed on the sward, it is frequently observed in this situation:—this has been generally ascribed to its specific gravity, and to its acting in a mechanical manner. In gravelly, or sandy soils, there can be no doubt but that the diffusibility and smallness of the particles of lime will induce it mechanically to sink through the larger particles of the sand or gravel, and to remain at rest on the more compact stratum which may resist its passage.

When lands of this description have been limed, and kept constantly under annual crops, the greater mechanical process of the plough will operate against the lesser one of subsidence, and keep the lime diffused through the soil: but in clayey or loamy soils, which are equally

diffusible

diffusible with lime, and nearly of the same specific gravity, the tendency which lime has to sink downwards, cannot be accounted for simply on mechanical principles.

In lands of this description, under the plough, the lime is dispersed or mixed with the soil, until such time as these lands are laid down with grass seeds. After re-maining in this situation at rest for a certain number of years, on breaking up, a floor of calcareous matter will frequently be found lying immediately beneath the roots of the grass. This effect, contrarily to the general opinion of its being disserviceable, is of great utility, as the staple or depth of the soil is always increased and rendered less retentive of water in proportion to the distance which the lime penetrates downwards ; and thus by increasing the depth of the soil a greater scope is afforded for the ex-pansion of the roots and nourishment of vegetables. These effects of lime in soils, except in those that are gravelly or sandy, cannot be accounted for simply on mechanical principles, but may probably be explained on such as are chemical.

Lime

Lime is capable of being dissolved in water, in six hundred times its own weight. Chalk, *i. e.* lime combined to a certain degree with fixable air, is insoluble; but chalk is capable of solution by a greater proportion of fixable air, either in consequence of its disengagement from vegetable matters decaying in the soil, or from the changes which the carbonaceous matter therein contained may undergo. When chalk in this manner is rendered soluble, the solution will sink through the surface mould to a more compact stratum, through which it cannot pass: in this situation it will lose the superabundant proportion of fixable air by which it previously had become soluble, and will again return to the state of chalk, forming a stratum infinitely more pure and unadulterated than could possibly have been formed, had the process been merely mechanical: in which case, contrary to the fact, all the argillaceous, magnesian, siliceous, and ferruginous matters, of equal degree of attenuation or size of parts, would have sunk with the lime. *

The

* Chalk and clay are nearly of the same specific gravity, and sand of greater specific gravity than lime.

The abundant use of lime has undoubtedly occasioned the consumption of a large proportion of the vegetable and animal matters in the soils to which it has repeatedly been applied; still the evil is not so great as at first it may appear. For although a considerable proportion of these matters has been disengaged in a gassious state, or otherways made to contribute to the growth of plants, still the lime, or chalk, by entering into certain combinations, has formed and accumulated a large stock of insoluble matter, capable, by certain saline substances, of being again brought into action, and of being rendered conducive to vegetation.

ALKA-

## ALKALINE SALTS

ARE fixed, or volatile. The fixed are obtained by the combustion of marine and terrene plants. Mineral alkali is also to be procured from muriat of soda, or sea salt. The volatile is obtained from urine—animal matter—fossile coal—soot—and other substances. Alkaline salts combine with acids, forming neutral salts; with animal fat, and oil, and vegetable oils, they form soap, or a saponaceous matter, diffusible in water. Alkaline salts act upon and destroy the continuity of the parts of animal and vegetable substances : they act most powerfully on the latter, when oxygenated, forming therewith saline compounds, which, in a very high degree, promote vegetation.

Oxygenated vegetable matter, and oxygenated peat, when decomposed, and rendered soluble by alkaline salts, assume a brownish red colour, and are tasteless: hence the alkali must enter into combination, and be neutralized by an acid or acids. These will be found to

G                                                      be

be the phosphoric and soreline, or, as it is now generally termed, the oxalic acid, forming, according to the particular alkali used, phosphat and oxalat of pot-ash — phosphat and oxalat of soda, or mineral alkali—phosphat and oxalat of ammoniac, or volatile alkali.

By exsiccation, the above-mentioned extract (which is very similar in its colour and effects on ground to the juice of dunghills) assumes the appearance of a darkish brown gum, soluble at any time by the application of water. Alkaline substances act in the same manner on oxygenated fossile coal, as they do on oxygenated vegetable matters or peat ; forming likewise a brownish red liquor, which equally promotes vegetation. The acids contained in oxygenated fossile coal, in a state of combination with calcareous matter, will probably be found to be the phosphoric acid, the acid of borax, and that of sorel, or the oxalic acid.

MAG-

## MAGNESIA, CONSIDERED AS A MANURE,

Is contained in steatities or soap rock, and in a variety of other earths and stones. It combines with acids, forming neutral salts, all of which are very soluble, and the greater part of them promote, in a very considerable degree, the growth of plants. Magnesian earths may be applied with peculiar advantage to soils generally, and not improperly called sour soils, containing green vitriol, arising from the decomposition of pyrites. It will decompose the metallic salt by superior affinity, and form with the acid, Epsom salt, known in a high degree to promote vegetation ; while the earth of iron will be separated in the state of an ochre, or iron combined with fixable air.

IRON.

## IRON.

THE metallic substance contained in the greatest quantity in soils, is iron; it exists therein in a variety of states—in a metallic state—in the state of yellow ochre, or iron combined with fixable air—in the state of red ochre, or iron combined with pure air or oxygen—and in the state of pyrites, or iron mineralized by sulphur. Pyrites are only found in such soils as have not long been under cultivation, or exposed to the action of the air for a sufficient length of time, so as to decompose the pyrites by the combination of pure air; a process in this Treatise called oxygenation. By this means, the vitriolic acid is formed, and which, as it forms, joins with the earth of iron, the other constituent part of the pyrites, and forms green vitriol. This salt, in poor soils, containing little calcareous matter, is extremely injurious to the growth of plants, the bad effects of which to vegetation may be corrected by dung, and urine of cattle—by lime—chalk—magnesia—and alkaline salts. With lime or chalk, the acid of the green vitriol

forms

forms gypsum, a salt which is very insoluble; whilst with magnesia and alkalis it forms Epsom salt, and vitriolated tartar, or Glauber salts: salts, whose beneficial effects on the growth of plants have been fully ascertained.

The proportion of iron in most soils, is so very considerable, that there can be no doubt but it was placed there, by the Great Creator of all things, to answer some wise purpose; and, although it is only found in small quantities in vegetable and animal substances, still its effects in promoting vegetation, may, in a chemical point of view, be much greater than can possibly be accounted for by the very small proportion of iron found in organic bodies.

The following is an attempt to explain how it may conduce to vegetation.

Iron, as has been observed, is found in soils under different modifications, changing from one state to another: when in that of pyrites it is changed, by the absorption of pure air or oxygen, to that of a metallic salt.
When

When alkalis, calcareous matter, or magnesia, are added to soils containing such metallic salts, they are decomposed; the acid combining with these substances in preference to the earth of iron, which is separated in the state of yellow ochre, or iron combined with fixable air.

By the addition of animal, vegetable substances, or carbonaceous matter, to lands containing iron in this state, it is brought towards a metallic state, and fixable air is disengaged. Ground, containing much iron in the state of ochre, is of a yellowish tinge, or colour; but as it receives dung or inflammable matter, the soil, together with the iron contained in it, will change to a brown or dark colou .

Iron thus reduced to a metallic state is capable of combining with pure air or oxygen, in which case fixable air would be generated by the combination thereof with the carbonaceous matter of the iron. The fixable air thus generated, may be disengaged from the earth of iron by a fresh application of inflammable matters. Nothing is at rest in this world; all bodies are undergoing successive changes from one state to another.

I                                                      On

On iron the changes are produced by the alternate action of inflammable matter, and pure air or oxygen : in the former case fixable air is disengaged; in the latter, atmospheric air is separated, or decomposed, by the abstraction from it of pure air. Both the azotic or phlogisticated air, the other component part of atmospheric air, and fixable air, promote the growth of plants. ˙

-------------------

## VITRIOLIC ACID.

THIS acid, in the new nomenclature, is called by the French chemists the sulphuric acid ; a name much more descriptive of its origin than that generally used.

All acids, in the new nomenclature, are named from the peculiar bases or substances of which they are formed, by the combination of pure air, or oxygen ; the presence of which in all cases is necessary to constitute an acid. This process, under the head of oxygenation, has been frequently resorted to, and which, together with the insoluble saline compounds formed by the developement

of

of certain acids from vegetable substances, and subsequent combination of them with calcareous matter, constitute the most prominent feature in the present work.

By a due knowledge and attention to this very important process of nature, the most beneficial consequences may be derived. It being a process, which, as it respects agriculture, has not been noticed by any writer on that science.

The vitriolic is the most powerful of all the acids. It disengages or expels other acids, when in a state of combination with metallic, earthy, or alkaline substances. When concentrated, it acts in a similar manner to that of alkaline salts, in the resolution or destruction of vegetable and animal substances, disengaging from them certain gasses, and forming therewith certain saponaceous and saline compounds. These solutions or extracts are of a reddish brown colour, similar to that produced by the action of alkaline salts on oxygenated peat. On the principles already stated, the vitriolic acid may be used beneficially to decompose, and to bring into action

the

the insoluble matter accumulated in soils, by the combination of the phosphoric and soreline acids with calcareous matter. The vitriolic acid will join with the calcareous matter, and form gypsum; whilst the posphoric and soreline acids, in consequence of their disengagement, will combine with other matters in the soil, particularly with magnesia, forming salts which are very soluble, and conducive to the growth of plants.

But the endless series of processes adopted by nature, do not finish here; for, on a supposition, that the phosphoric and soreline acids had been fully disengaged from the calcareous matter, with which they had formerly been united, and that in the state of phosphat, and oxalat of potash, soda, ammoniac, or magnesia, they had expended themselves in the process of vegetation; still the gypsum remaining in the soil would, on a renewed application of dung, animal or vegetable matter, be brought from the state of gypsum, which is insoluble, to a state approaching to that of a hepar of lime, which is soluble; and as the vitriolic acid and calcareous matter are contained in, and form a part of the compounded residuum

H                                              of

of vegetable matters, it may hence be inferred, that
these matters were not generated in, but were taken up,
when in a state of solution, by the roots of plants. Thus
may the good effects of gypsum in America be accounted
for. To these beneficial effects, from the combination of
inflammable substances with gypsum, forming what is
called a hepar or liver of sulphur, may be added the
large share of nourishment, which trefoils, and plants of
a certain formation of stem and leaf, receive, by the
hepatic air disengaged from hepars, when they, by the
process of oxygenation, are again returning to the state
of neutral salts, of which such hepars had been formed
by the combination of inflammable or carbonaceous
matter.

Thus by a thorough knowledge and application of
chemistry to agriculture, the several substances in soils
may be made to undergo a varied change of new combinations, tending to promote the greatest of all objects,
a more plentiful supply of food.

NITROUS

## NITROUS ACID.

NITROUS acid is a compound of phlogisticated air or azotic gas with pure air or oxygen. This combination takes place under different circumstances, particularly by the putrefaction and decay of animal and vegetable substances. As it forms, it combines either with calcareous matter or alkaline salts; forming saline substances, which are conducive to vegetation.

The nitrous acid, in the table of affinities, is placed next to that of the vitriolic.

---

## MARINE, OR MURIATIC ACID.

THIS acid is next in affinity to the nitrous, and constitutes about one third of muriat of soda, or sea salt. It consists, like all other acids, of a combination of pure air or oxygen with a peculiar basis, which has not as yet been

been ascertained, but is supposed to be of an inflammable nature. Marine acid is contained in sea water, in salt rock, and in salt springs, combined with mineral alkali, or soda, and with the earth of magnesia.

The disengagement and separation of·this acid, from the alkaline basis with which it is united in sea or rock salt, may be accomplished by various methods; one only has as yet been discovered and effected at an expence which can admit a manufactory of alkaline salts being established on an extensive scale. The accomplishment of this most desirable object, by a cheap and easy process, must appear, with respect to certain useful arts, as well as to the application of it to agriculture, to be one of the most important discoveries to which chemistry could have lent its aid.

PHOS-

## PHOSPHORIC ACID, AND SORELINE OR OXALIC ACID.

THE origin, nature, and properties of these acids have previously been so fully discussed, that any further explanation is deemed unnecessary.

---

## INSECTS.

INDEPENDENT of the various substances hitherto noticed, as being contained in soils, and affording their assistance in the production and nourishment of plants, the innumerable tribes of insects which abound in rich soils, or soils long under cultivation, ought not to be overlooked. Some of these insects are extremely noxious, whilst others·are inoffensive to the vegetable kingdom; but all, when destroyed, (as will hereafter directed) may be rendered serviceable in affording

T                                                     a proa proportion of animal matter for the general uses of vegetation. The common earth worm, which is inoffensive, may be made to rise to the surface, and to become useful for the domestic purpose of feeding poultry.

------

## SALINE SUBSTANCES WITH EARTHY BASES.

Pure clay, chalk or calcareous matter, magnesia, and the earth of iron, are capable of being dissolved by many acids, and of forming salts, which are more or less soluble. *

As some one or more of these salts are contained in most soils, it is an object of great importance to agriculture, to ascertain what are the combinations which different acids make with the earths above-mentioned; which of them are beneficial, and which injurious to vegetation ; as also how such injurious effects may be corrected.

The

* Siliceous matter is not included, being soluble only by the fluor acid.

The following table will shew, the salts that may be formed, by the combination of the vitriolic—nitrous—muriatic—phosphoric—and soreline or oxalic acids, with clay, chalk, magnesia, and the earth of iron. To this a discussion more at large will be added on the respective properties of the several combinations thus formed, in promoting, or in retarding vegetation, agreeably to the following order.

## TABLE.

| | | |
|---|---|---|
| Clay — Vitriolic or Sulphuric acid | Sulphat of Argill or Allum. |
| Chalk — — do. — | Sulphat of Lime or Gypsum. |
| Magnesia — do. — | Do. of Magnesia or Epsom Salt. |
| Earth of Iron — do. — | Sulphat of Iron or Green Vitriol. |
| Chalk — · — Nitrous acid — | Nitrat of Lime. |
| Magnesia — do. — | Nitrat of Magnesia. |
| Earth of Iron — do. — | Nitrat of Iron. |
| Chalk Marine or Muriatic acid | Muriat of Lime. |
| Magnesia — — do. — | Muriat of Magnesia. |
| Earth of Iron — do. — | Muriat of Iron. |
| Chalk — Phosphoric acid — | Phosphat of Lime. |
| Magnesia — — do. — | Phosphat of Magnesia. |
| Earth of Iron — do. — | Phosphat of Iron or Sederite. |
| Chalk — Soreline or Oxalic acid | Oxalat of Lime. |
| Magnesia — do. — | Oxalat of Magnesia. |
| Earth of Iron — do. — | Oxalat of Iron. |

## SULPHAT OF ARGIL, OR ALLUM.

ALLUM is a salt dissolvable in fifteen times its weight of water. It is produced in great quantities, by the decomposition of alluminous schyst or slate, on exposure to air, or by calcination.

Dr. FRANCIS HOME [*] of Edinburgh, the first person who thought of making experiments with saline bodies in promoting the growth of plants, found no beneficial effects to result from the application of allum to garden mould, the soil on which his experiments were made. Allum is contained in many soils, and is daily forming by the decomposition of alluminous schyst.

Where found in abundance, the soil is very properly denominated, by country people, a sour soil, on which few vegetables will grow. This sterility is to be corrected by lime, by earthy matters containing magnesia, and by

I
alka-

---

[*] Vide Dr. HOME's most valuable Treatise on that subject, published at Edinburgh in 1756.

alkaline salts. The neutral salts formed by such applications will be gypsum, Epsom salt, sulphat of potash, sulphat of soda, and sulphat of ammoniac, according to the
species of alkali applied. Although no beneficial effects
were found to result from the experiments made by Dr.
HOME, yet they may with great probability be expected
to arise by the application of allum to soils containing much calcareous matter; especially to such as
contain, beside this latter substance, a sufficient proportion of animal and vegetable matter. The allum will
in this case be decomposed by the calcareous matter, on
the principle of superior affinity, whilst the fixable air of
the calcareous matter will be disengaged, and on being
absorbed by the roots of plants, will afford them the carbonaceous principle.

SULPHAT

## SULPHAT OF LIME, OR GYPSUM.

Gypsum exists in great abundance, in many soils. It is prodeced by the decomposition of alluminous schyst, containing a due proportion of calcareous matter; with which the sulphuric acid will join, as it is formed in preference to the earth of allum or clay. It is likewise formed by the decomposition of pyrites, in such soils as contain a sufficiency of calcareous matter for the sulphuric acid to combine with, in preference to the earth of iron, the other constituent part of pyrites: and it is found in immense quantities, constituting not only the soil, but the substratum, of some countries, to a great depth. Gypsum is to be decomposed by alkaline salts; the sulphuric acid forming with them sulphat of potash and sulphat of soda, according to the alkali used. It is a salt very insoluble, requiring upwards of five hundred times its weight of water to dissolve it: hence supposing it equally deleterious to vegetation, as allum hath been considered, which is soluble in only fifteen times its weight of water, it must prove less injurious,

I 2

rious, from its greater degree of insolubility : but gypsum, far from being hurtful to vegetation when applied
to certain soils, promotes vegetation in a very high degree,
as is evinced by the use of it in some parts of the Continent of Europe and of America; and is further proved by
the chemical analysis of vegetables, whose ashes are
found to contain a certain portion of the component parts
of gypsum.  For the particular explanation of its mode
of acting on soils suited to it—See Vitriolic Acid.

----

## SULPHAT OF MAGNESIA, OR EPSOM SALT.

SULPHAT of magnesia, or Epsom salt, is not to be
found in such abundance as allum or gypsum: it is contained in sea water, and is to be procured in greater quantities by decomposing the muriat of magnesia, or the
bitter refuse liquor of salt works, by the means of the
vitriolic acid.  Native Epsom salt is sometimes to be met
with in the mineral strata, mixed with clay and siliceous
matter.  It is formed in soils containing alluminous
schyst, or pyrites, and a due proportion of the earth of

2                                    magne-

magnesia: in which case, on the decomposition of the schyst, or pyrites, the vitriolic acid so formed, will, by superior affinity, join with the magnesia, and form Epsom salt, in preference to the argillaceous or ferruginous matter of the above-mentioned substances.

Epsom salt is very soluble, dissolving in about twice its weight of cold water. It is decomposed by lime and alkalis, forming therewith gypsum, vitriolated tartar, Glauber salt, or vitriolic ammoniac. By Dr. HOME's experiments, Epsom salt has been found in a very high degree to promote vegetation. He states, that it made the garden mould he used for his experiments produce one fourth more grain.

SULPHAT

## SULPHAT OF IRON, OR GREEN VITRIOL.

THIS salt is formed naturally, in many places in great abundance, by the process of oxygenation, from sulphureous or pyriteous substances. These matters are generally found accompanying the coal strata, as well as in coal itself; particularly in such coals as are sulphureous. This salt is very soluble in water, and is in a high degree injurious to vegetation, when it abounds in soils consisting of poor clay and siliceous matter, without any admixture of vegetable or calcareous substances.——— It is decomposed by alkaline salts, forming therewith vitriolated tartar, Glauber salt, vitriolic ammoniac, gypsum, and Epsom salt. When added to soils containing calcareous matter, and a due proportion of animal and vegetable substances, it has been found, when not used in too great quantities, to have produced beneficial effects in promoting the growth of grass ; but experiments have not as yet been made fully to ascertain its effects on arable land.

NITRAT

## NITRAT OF LIME.

THIS saline substance is found in the rubbish of old buildings, and in those materials from which salt-petre is extracted : viz. animal and vegetable matters, which, with a due proportion of calcareous earth, have undergone the putrefactive process, together with a subsequent, sufficiently long, exposure to atmospheric air. According to Dr. HOME, it is likewise contained in what is commonly called hard water, which, by his experiments, was found to promote the growth of plants in a much higher degree than soft water.

Nitrat of lime is very soluble, and is deliquescent; it is decomposed by fixed alkalis, and forms therewith nitrat of potash or salt-petre, and nitrat of soda, or cubic nitre.

## NITRAT OF MAGNESIA.

On this subject, with precision, little can be advanced, as no agricultural experiments have, as yet, been made with this compound; it is a very deliquescent and soluble salt; and there is reason to expect, that it will produce effects, in promoting vegetation, similar to those which may result from the application of the nitrat of lime.

It is decomposed by alkalis and lime, and forms there-with nitrat of potash, nitrat of soda, and nitrat of lime.

---

## NITRAT OF IRON.

As this salt is rarely presented by nature, its properties or effects, as they may apply to Agriculture, are not worthy of observation.

MURIAT

## MURIAT OF LIME.

THIS salt is found in a small proportion, in sea water. It is very soluble, and when mixed with dung, its effects in promoting vegetation will probably be found similar to those of the next article, the muriat of magnesia. It is decomposed by fixed alkalis, forming therewith muriat of potash, or digestive salt of silvius, and muriat of soda, or sea-salt.

## MURIAT OF MAGNESIA

Is found in great abundance in sea water, constituting upwards of one fourth of the saline matter it contains. It may be procured in great quantities from the bitter refuse liquor which at present runs to waste at the salt works.

It is a salt very deliquescent, and of difficult chrystallization; its acid is capable, in a great measure, of being

K                              expelled

expelled by heat, and very considerable benefit has been experienced from its use, in promoting vegetation, when mixed with dung, or compost dung-hills. It seems to possess, when applied in moderate quantities, the septic powers of sea salt, and thus to promote the complete putrefaction of dung.

It is decomposed by fixed alkalis and lime, forming therewith digestive salt of silvius, sea salt, and muriat of lime. With ammoniac, or volatile alkali, it forms a triplicate salt of easy chrystallization.

---

## MURIAT OF IRON.

THIS salt seldom occurs in nature. It is highly inimical to vegetation. It is decomposed by fixed alkalis, lime, and magnesia, forming digestive salt, sea salt, muriat of lime, and muriat of magnesia.

PHOS-

## PHOSPHAT OF LIME

Is contained in animal matters, such as bones, urine, shells, &c. &c. in some sorts of limestone, and in vegetable substances, particularly in the gluten, or vegeto-animal matter of wheat or other grain. It is a saline compound very insoluble. There is reason to believe, a very considerable proportion of this nearly insoluble salt is contained in most fertile soils, especially those that have been long under cultivation. It is not to be decomposed by pure alkalis; but this may be effected by mild vegetable and mineral alkalis, on the principle of the double elective attractions; in which case, carbonat of lime (or chalk) will be precipitated, and the phosphoric acid will join with the alkali, and form phosphat of potash, or phosphat of soda, according to the alkali applied. These alkaline phosphats will be found to promote vegetation in a very great degree: the substances of which they are composed, viz. alkaline salts and phosphoric acid, are found in the ashes of most vegetables.

PHOS-

## PHOSPHAT OF MAGNESIA.

This is a very soluble salt; seldom occurs in nature; promotes vegetation. It may be formed in soils containing phosphat of lime and uncombined magnesia, by watering ground, containing these substances, with water acidulated by the vitriolic acid, or by an acid pyriteous liquor; or the acid may be applied, by moistening mould with water properly acidulated by the vitriolic acid, and then sowing, or spreading the mould on the ground. In this case the vitriolic acid will join with the calcareous matter of the phosphat of lime, and form gypsum; whilst the phosphoric acid, thus disengaged, will join with the magnesia in the soil, and form phosphat of magnesia.

If phosphat of lime can be decomposed by sulphat of iron, (on the principles of the double electrive attractions) and form sulphat of lime (or gypsum) and phosphat of iron, the phosphat of iron may, in that case, be decomposed by magnesia, forming phosphat of magnesia.

PHOS-

## PHOSPHAT OF IRON

Is to be decomposed by alkaline salts and magnesia.

---

## OXALAT OF LIME.

Oxalat of lime is a very insoluble saline compound, formed by the combination of calcareous matter with the oxalic, or soreline acid, with which acid calcareous matter has a greater affinity than have alkaline salts. The oxalic acid, or acid of sorel, is found in the plant of the same name, combined with the vegetable alkali, in the state of a superacidulated salt; and it is likewise contained in many other vegetables. Oxalat of lime is to be produced, by adding calcareous matter to the expressed juice, or salt of sorel. The calcareous matter will combine with the acid, and form oxalat of lime; whilst

whilst the carbonic acid, or fixable air, will combine with the vegetable alkali, forming mild alkali, or carbonat of potash.

Much of the oxalat of lime will be formed, by adding calcareous matter to ground abounding with plants of sorel, or other vegetables containing the soreline or oxalic acid; by which application, the vegetable alkali will be disengaged in a carbonated or mild state. From the insolubility of oxalat of lime, it is not probable that it can contribute, *per se*, to the food of plants. It cannot be decomposed by alkalis, on superior affinity, because its affinity is greater with calcareous matter; but it may be decomposed by the vitriolic acid, in which case, gypsum will be formed, and the soreline, or oxalic acid, thus disengaged, will be capable of entering into new combinations with fixed, or volatile alkaline salts, or magnesia. These combinations are soluble, and, when not superacidulated, promote vegetation in a high degree.

Oxalat of lime may probably be decomposed by certain neutral salts, on the principles of the double elec-

electrive attractions, especially if the neutral salts, used for that purpose, are superacidulated.

Oxalat of lime is decomposed by fire; the acid is destroyed, and carbonic acid, or fixable air, is formed, and disengaged.

---

## OXALAT OF MAGNESIA.

THIS is a soluble salt, and, when not superacidulated, promotes vegetation. It is formed, by adding magnesia, or earthy matters, containing magnesia, to the juice, or salt of sorel; the superabundant acid of which will combine with the magnesia. It is probable, that that part of the juice of sorel, which is in a neutral state, will be decomposed by the greater affinity that magnesia has to the soreline or oxalic acid; and that the vegetable alkali will be disengaged in a mild state, in the same manner as if calcareous matter, or chalk, had been used.

OXALAT

## OXALAT OF IRON.

Nothing certain can be advanced on the subject of this salt, farther than that it is decomposed by lime, magnesia, and alkalis.

---

## SALINE SUBSTANCES COMPOSED OF ALKALIS AND ACIDS.

The nature of the simple earths, and the compounds they form with different acids, having been explained, it will now be proper to give a list of such neutral salts as may be formed by the combination of the acids already mentioned with alkaline salts.

These are stated in the following table, to which will be added, under each distinct head, the peculiar powers of each, in promoting or retarding vegetation, &c.

TABLE

# TABLE.

| Potash | Combined with Vitriolic or Sulphuric Acid | Sulphat of Potash, or Vitriolated Tartar |
|---|---|---|
| Soda | Ditto | Sulphat of Soda, or Glauber Salts |
| Ammoniac | Ditto | Ditto of Ammoniac, or Vitriolated Ammoniac |
| Potash | Nitrous Acid | Nitrat of Potash, or Salt-Petre |
| Soda | Ditto | Ditto of Soda, or Cubic Nitre |
| Ammoniac | Ditto | Ditto of Ammoniac |
| Potash | Marine Acid | Muriat of Potash, or Digestive Salt of Silvius |
| Soda | Ditto | Ditto of Soda, or Sea Salt |
| Ammoniac | Ditto | Ditto of Ammoniac, or Sal Ammoniac |
| Potash | Phosphoric Acid | Phosphat of Potash |
| Soda | Ditto | Ditto of Soda |
| Ammoniac | Ditto | Ditto of Ammoniac |
| Potash | Soreline, or Oxalic Acid | Oxalat of Potash |
| Soda | Ditto | Ditto of Soda |
| Ammoniac | Ditto | Ditto of Ammoniac |

L                                               SUL-

## SULPHAT OF POTASH, OR VITRIOLATED TARTAR.

THIS salt is soluble in fifteen times its weight of cold water, and has been found by Dr. HOME to promote vegetation in an extraordinary manner. The garden mould on which his experiments were made produced an increase of one-fourth more grain, in consequence of the application of this salt. Vitriolated tartar is to be had from most vegetable matters by combustion; it forms at least one-third of the saline matter obtained by the lixiviation of their ashes. This is sufficient, *a priori*, to prove, (independently of Dr. HOME's very satisfactory experiments) that vitriolated tartar is beneficial to vegetation. This substance is a refuse article in some branches of manufacture; but the quantity produced is a mere trifle, in comparison to the quantity that might advantageously be applied to the purposes of Agriculture.


SULPHAT

## SULPHAT OF SODA, OR GLAUBER SALT.

GLAUBER Salt is soluble in about two and one half times its weight of cold water. From experiments, it has been proved to promote vegetation in a very high degree. It is procured in small quantities, in the processes for making the muriatic acid, and muriat of ammoniac, or sal ammoniac. The high price at present of this article precludes the use of it; but could it be made and sold at a cheap rate, it would prove a most valuable acquisition to Agriculture.

---

## SULPHAT OF AMMONIAC OR VITRIOLIC AMMONIAC.

THIS salt is very soluble, and promotes vegetation; but it is not to be had in such quantities as to render it an article of importance to Agriculture.

## NITRAT OF POTASH, OR SALT-PETRE,

Is a salt very soluble, formed by the putrefaction and complete decomposition of animal and vegetable substances, when mixed with calcareous matter and wood ashes. It promotes vegetation, but its high price precludes its being used as an article of manure.

---

## NITRAT OF SODA, OR CUBIC NITRE.

This salt is likewise very soluble, but seldom naturally occurs. It is probable, that cubic nitre would promote vegetation in an equal degree with nitrat of potash or salt-petre.

NITRAT

## NITRAT OF AMMONIAC.

As this is a salt that cannot be procured in sufficient quantities, for the purposes of Agriculture, nothing need be observed on its nature or properties.

---

## MURIAT OF POTASH, OR DIGESTIVE SALT OF SILVIUS.

DigestivE salt of Silvius is soluble in about thrice its weight of cold water. As this salt does not often naturally present itself, and cannot be procured in large quantities, its effects on vegetation may be considered to be stated under the next article, which it very much resembles.

MURIAT

## MURIAT OF SODA, OR SEA SALT.

This salt is soluble in about three and a half times its weight of cold water, and is to be had in unlimited quantities from sea-water, salt springs, and salt mines. It never yet has been fully ascertained, whether it is beneficial or injurious to vegetation. Its fertilizing powers have been highly extolled by some, whilst others have positively denied their efficacy. It is probable, that any power it may possess, in promoting vegetation, depends on its septic quality, or power it has of assisting putrefaction, when mixed in small or due proportions with dung or vegetable substances; though in large quantities it is antiseptic, and prevents putrefaction. *

These different effects of sea salt are the best explanation that can be given of the several opinions formed of this article as a manure, which, applied in small quantities, may prove beneficial to certain soils; whilst, in large quantities, it must be injurious to all.

Exclu-

* Vide Sir John Pringll's Essays.

Exclusively of its septic effects, when used with dung, and animal or vegetable matters, it is destructive to many different kinds of living insects, such as snails, grubs, slugs, worms, &c. Sea salt is contained in great quantities in marine plants; and as there is a variety of plants approaching more or less to those of a marine nature, its effects will be found greater in promoting the growth of certain classes of vegetables than of others. Sea salt is extremely injurious to poor soils, and ought only to be applied to rich lands.

## MURIAT OF AMMONIAC, OR SAL AMMONIAC.

NOTHING need be said on this subject, as the expence of the article precludes the use of it in Agriculture.

## PHOSPHAT OF POTASH—OF SODA—AND OF AMMONIAC.

THE above salts having similar effects in promoting vegetation, the same description, as it respects Agriculture, will nearly apply to them all.

They

They are found in the greatest quantities in urine; the good effects of which, on most soils, being well known, it may be presumed that similar salts to those contained therein must be equally efficacious.

Phosphoric acid and alkaline salts are contained in the ashes of vegetables; these salts, or their primary principles, must necessarily have constituted a part of such vegetables, from which it may be inferred, *a priori*, that alkaline phosphats are conducive to the growth of plants. Alkaline phosphats are to be obtained by the addition of alkaline carbonats, or mild alkaline salts, to oxygenated peat, or other oxygenated vegetable substances, forming therewith a reddish brown mucilaginous compound.

----

## OXALAT OF POTASH, OF SODA, AND OF AMMONIAC.

THESE soluble salts very highly promote vegetation; and may be had in great abundance by the addition of alkaline salts and other saline matters to oxygenated peat, and also to oxygenated fossile coal, forming there-

with

with a mucilaginous saponaceous compound, soluble in water, similar to that mentioned under the last article.

The action of alkaline salts here stated to take place on oxygenated peat, a substance which previously has been observed to consist of the phosphoric and soreline, or oxalic acids, combined with calcareous matter, is undoubtedly, as it may respect the action of alkalis on the oxalat of lime, contrary to chemical principles, as lime has a greater affinity with the soreline or oxalic acid than alkalis have; but as peat, in the state here described, is mixed with other saline matters, and does not in all respects resemble the oxalat of lime, but contains also a large proportion of carbonaceous and inflammable matter, the action of the alkali, in dissolving the peat, must be ascribed to the combination of the inflammable or other matters.

## STABLE, FARM YARD DUNG, AND COMPOSTS.

WHEN CATO was asked what was the best system of farming, he thrice answered, that the best system of farming was to procure food for cattle; a reply which refers to obvious consequences, and requires no explanation.

Although it is a common saying in Scotland, that " muck is the mother of the meal chest," still there is no country where the preservation of the urine of cattle, and the juices of dung-heaps, are so little attended to ; or where the farmers, in general, are at so little pains to procure the greatest quantity of an article so indispensably necessary to the obtaining abundant crops.

In England, the farmers are much more attentive to the making and collecting this species of manure, and their conveniencies of farm-yards, and other places for preserving dung, are better chosen. These are not only kept fully littered, but the lanes and hollow ways, in the vicinity

vicinity of the farm-yard, are frequently filled with haulm, or inferior straw. The putrefaction of it is promoted by being trodden, and by receiving the urine of passing cattle, which would be much facilitated, were the place in which it is deposited overshaded with trees, and sheltered from the too great action of wind and rain.

In Scotland, a preference is generally given to a sloping bank or rising ground for the situation of farmhouses and offices; and, as there are but few instances, unless of a recent date, where an inclosed farm-yard is to be met with, the urine from the stables and cowhouses, as well as the juices from the dung heap, run off, or are washed away by the rains, and turn to no good account.

The want of proper farm-yards and conveniencies is but one of the reasons why a less quantity of dung is made by the farmers in Scotland than by those of England. The principal one is, that as the greater part of Scotland is better adapted to the breeding of cattle than to the production of grain, the whole of the straw pro-

duced

duced from the arable land of each farm is eaten in the succeeding winter by the stock; and, as few farmers have more than enough for their cattle, little or none can be afforded as litter for their stables and farm-yards. In severe winters, of long continuance, a scarcity of straw for their cattle is frequently experienced. This proves that most farmers in Scotland keep a greater number of cattle during summer than they ought prudently to calculate on maintaining through the winter. It is very unusual in Scotland to grow hay but for the purpose of sale. The working horses are fed with oat straw until seed time; when, as an article of luxury, they are supplied with pease or bean straw, the barley straw being allotted to the cows and young stock.

There are many who rent small farms, still more œconomical in the consumption of their straw and fodder. The whole of the straw, of every kind, is by them given to their horses; whilst their cows, and neat cattle, are exclusively fed with the coarsest parts of the straw rejected by the horses, broken and bruised by their feet, and well drenched by their urine; a sort of food which does not appear very palatable, and which, although contrary

trary to the first impression it makes, is certainly not to be attributed alone to scarcity, but to choice, as cows and neat cattle are frequently observed to forsake the best provender, and to pick up and eat the litter from horse-stables; the state of the stomachs of these animals inducing them to seek and apply, as a stimulus, the volatile alkali contained in the urine of the horses, absorbed by the straw. The necessity that cattle are under of thus supplying themselves with the only stimulus or saline matter within their reach, and the well known salutary effects of salt, when given to cattle in other countries, evidently point out the very important benefits that would arise, were cattle supplied from time to time with a due proportion of sea salt; and, as such, it cannot but be regretted, that the duty on an article, so essential to the purposes of farming and grazing, should so completely operate as a prohibition to its use : and the more so, as Government has long since been in possession of suggestions which could not have failed, had they been duly attended to, to have insured the most important benefits, not only to agriculture, the feeding and health of cattle, but likewise to several branches of manufacture in

these

these kingdoms, without a probability of causing any diminution in the public revenue.

Did the conveniencies attached to a Scotch farm allow the industry of the tenant to be exerted in preserving the urine and dung of his cattle, by constantly bedding them with, or mixing their urine and dung with dry peat, or when this substance is not to be procured, with rich black mould, the consumption of all the straw by cattle would, in such case, be found to be a practice highly conducive to the interest of the farmer.

Mr. Bakewell and Mr. Chaplain, two very skilful breeders of stock in England, are said to pursue very different modes in the feeding and maintenance of their horses and cattle. The one gentleman is said to supply them with hay, and to use all his straw for litter; whilst the other consumes the whole produce of hay and straw from his farm in feeding and maintaining of cattle. A platform is constructed on which the cattle stand, sufficiently open in the seams to allow them being kept clean and dry, without any bedding of straw. This method admits
mits

mits a greater number of cattle to be maintained, and a greater quantity of *real* dung to be procured, than when a less number of stock are kept and well littered.

Food in its passage through the bodies of animals becomes mixed with animalized matter, and consequently more rich and more valuable, weight for weight, as a manure, than dung procured by littering cattle, although there must necessarily be much less in bulk or quantity, from the large proportion of the food of animals, which goes off by breathing and insensible perspiration; beside which, without the utmost care, it is extremely difficult to prevent the urine and the valuable juices of the dung from sinking through the floors of cow-houses and stables, or the soil of farm-yards. Could these inconveniencies be effectually provided against by a proper flooring of clay or chalk, a preference appears due to the consumption of the whole of the produce by cattle, provided that attention be paid to the mixing daily a sufficient quantity of peat or mould with the dung and urine, so as completely to absorb and take up whatever may remain of these matters in a fluid state. By this process there can be no doubt that a greater quantity, and a still more valuable

dung

dung may be obtained than by the other practice of keeping a less number of cattle, and littering them with straw.

When peat cannot be had, the richest and blackest mould should be procured, that the volatile alkali of the urine may act upon and dissolve, into a mucilaginous gummy liquor, the oxygenated inert vegetable matter contained in such mould. Peat, however, is to be preferred, because it contains a more abundant proportion of oxygenated vegetable matter for the volatile alkali to act upon.

These are not to be considered as theoretical statements, but the result of actual experiments, attentively made in Scotland. The quantity of manure made in the same given time was much greater than if litter had been used ; and the manure procured was infinitely more rich and valuable. These experiments were not confined to the dung and urine of cattle, but the chamber-lye of the family was carefully preserved, and mixed also with a due proportion of oxygenated peat, which was found to produce a greater effect in dissolving the peat than the uri ne from the cattle.

The

The importance of making or obtaining the greatest quantity of manure with the materials now generally known, and capable of being procured, is, in this practical Dissertation, made to precede the preparation of all other manures or composts of a more expensive nature. Prudence and œconomy point out, that what is easiest and cheapest to be done, should always be first done; and that recourse should not be had to other means, until that source of supply is exhausted.

In the former part of this Treatise, under the head of Oxygenation, a short remark was made, that stable dung, by long keeping, lost its more fertilizing saline parts, and became oxygenated, and insoluble. A heap of such dung, kept for some years, has been known to become inodorous, insoluble, and in all respects similar to, and was a true peat; hence the practical inference, that dung should only be kept a certain time.

When animal dung and vegetable matters are mixed together, such as horse dung, urine, straw, and hay, a degree of heat is generated and disengaged by the absorption of vital air, or oxygen, and water is decomposed.

N                                        As

As the process of putrefaction advances, ammoniac or volatile alkali, (a compound of inflammable and phlogisticated air) is formed, and, in its tendency to escape from the heap, combines with such parts of the vegetables and matters of the dung, as had advanced to the oxygenated state, forming therewith the saponaceous saline matter so often adverted to in this Treatise. The formation of this saponaceous matter, in the greatest possible quantity, will be promoted by mixing and covering the dung with a due proportion of earth; hence the dung of hotbeds is that which is most completely rotted, and most asimilated to the saline saponaceous state above described; and in this state is more capable of promoting vegetation, than dung that had not arrived to an equally advanced state of putrefaction.

Many farmers differ in opinion as to the propriety or the advantages which attend using long or fresh dung, or dung which is completely rotted. This disputed point seems capable of adjustment. Were the views of the farmer to promote only the next immediate crop of grass or grain, the dung, when applied, should be fully and completely rotted; but if his views extend to subsequent crops,

crops, or the soil be of a nature to receive benefit by the fermentation and heat produced by the application of fresh dung, preference should undoubtedly be given to dung in a long state, provided it is immediately ploughed in, and totally covered, which is not easily accomplished with dung of this description. Long dung is always to be preferred in the culture of potatoes; for dung completely rotted frequently causes this crop to be watery and worm eaten. Many farmers only apply coarse straw or litter; whence it might be imagined, that the benefit arising from such an application, must be more dependent on the straw mechanically keeping the ground open or loose, than in contributing, by any part of its own substance, to the growth of the potatoes, which cannot well be supposed; as the straw, in digging up the potatoes, is generally found in an undecayed state. It is highly probable, that the atmospheric air contained in the intervals of the soil, thus made by the straw, may suffer a degree of separation, or decomposition in its (as it were) imprisoned state, by which means the pure air or oxygen may combine with the straw, and inflammable or vegetable matter, in the soil; whilst the azotic or phlogisticated air will contribute to the growth

of

of the plants. This explication of the beneficial effects arising to vegetation by stagnated air, will also account for the benefit which plants of a certain construction of stem and leaf, and which very much overshadow and cover the ground, ultimately receive, by preventing a free circulation of air. The application of long or short dung to ground, must appear too material to the practical farmer to be overlooked. The preference, in most cases, is undoubtedly to be given to such dung as has most completely undergone the putrefactive process. Under this head it is necessary to notice, that dung and urine newly voided (unless when animals are diseased) are not in a putrescent state. The time of retention in the body of animals is of too short a continuance to allow that effect to take place. Such excrements are in a state advancing towards putridity, or in a small degree only putrid; a process which, to a certain extent, is necessary to stimulate the intestines to discharge the fæces.

The further putrescency of excrements is promoted by a due degree of heat and moisture, particularly when aided by certain saline matters. The most efficacious are neutral salts, containing the vitriolic acid; such as

vitrio-

vitriolated tartar, Glauber salt, Epsom salt, and gypsum. These neutral salts, by being mixed with putrescent sub- stances, are changed to the state of hepars : hence the extremely offensive and putrid effluvia disengaged from dung, or other putrescent matters, to which such salts had previously been added.

Similar effects, in promoting the putrefaction of dung, have not been experienced when muriat of soda, or pure sea salt, has been used ; and as no salt made in Bri- tain, whether from sea water, rock salt, or salt springs, is free from gypsum and Epsom salt, there is reason to suspect that the septic power of sea salt, when applied in small quantities (as stated by Sir JOHN PRINGLE and Dr. MACBRIDE) to animal and vegetable substances, may be owing in a great measure to the vitriolic salts contained therein, when not overpowered by too great a proportion of the muriat of soda, or pure sea salt, which is highly antiseptic, although considerably less so than other salts, such as. sugar and salt-petre.    Hence it is that sea salt of the greatest purity should exclusively be used for curing beef, pork, fish, cheese, and butter; whilst the more impure salt, *i. e.* such  as  contains

a great

a great proportion of vitriolic salts, should be exclusively applied to dung or other substances, to promote their complete putrefaction.

The purifying sea salt, or separating from it these other salts, would be attended with the two-fold advantage of improving the quality and taste of salted provisions, and of promoting the putrescency of dung, or other matters, by the application of such particular salts as are best adapted to their respective purposes.

A Treatise on the Purification of Salt, and the Necessity of an Alteration in the Present Salt Laws, was given to the Public about ten years since. The present system however is still continued, and individuals are content to make use of salt with all its impurities. The purifying of salt cannot take place, whilst the refuse bitter liquor, consisting of Epsom salt and muriat of magnesia, is not allowed to be removed from any salt work, or employed in any way whatever, without its becoming liable to a duty of twenty pounds per ton on the chrystallized salt capable of being thence obtained. This duty is equal to double that at present paid on sea salt, and amounts to a

complete

complete prohibition on its use, either as a medicine or as an article of manure. As a very large proportion of most sea salt consists of these bitter and septic salts, no salt manufacturer will be willing to incur a certain waste and loss, in the purification of salt ; whilst he cannot dispose of the refuse, and whilst the ignorance and inattention of the consumer occasions as high a price to be paid for impure salt, as if it were of the best quality. A revision and correction of the salt laws, at this time, would not only be extremely important to rural œconomies, but would be highly conducive to the health of our seamen.

These remarks on sea salt are of a nature not to be overlooked, and ought not to be deemed as too digressive from the subject of manures now under consideration. The procuring the greatest possible quantity of such matters, and in the highest state of preparation, should be the primary object of every farmer.

The dung naturally arising from the annual production of hay, straw and grain, and consumed by cattle or otherways, being in few instances sufficient to insure

abun-

abundant crops, from every part of a farm, every opportunity should be embraced by the skilful and intelligent
farmer to procure a further supply of stable, and other
dung, from large towns, as well as to provide himself
with other ubstances, which, as manures, he may apply
to improve, and bring into cultivation, his inferior lands.
As the dung which is naturally produced on every farm
must be limited, recourse should be had to other articles
of manure, for the purpose of promoting the highest
possible fertility of the soil, and thereby acquiring a
greater return to the dung heap for the succeeding
crops.   Such adventitious aids or helps to a farm, are of
a nature that answers a much better purpose as top-dressings to grass lands, than for lands which are constantly
kept under the plough.   The application of top-dressings
has, perhaps, been too little attended to, in consequence
of farmers being unacquainted with the resulting advantages to ground, when converted from pasture to
arable, by previously promoting the most luxuriant
growth of perennial grasses. By assisting the vegetation,
and increasing the vigour of perennial plants, their roots
are made to strike deeper down, and improve the staple
of the soil; with annual plants the same benefit is not to
be

be expected, as their growth and decay are limited to one season. Were manures exclusively applied, under a sys tem of convertible husbandry, to grass grounds, the lands would regularly be broken up, in due rotation of cropping; and there can be no doubt, but that a greater quantity of corn and herbage would annually be produced: and it is very probable that wheat and other grain would be less liable than at present to diseases, many of which, there is reason to believe, are occasioned by the immediate application of dung previous to sowing.

Top dressing, especially to meadow and pasture ground, is undoubtedly the best mode of applying manure. This practice is better understood in England than in any other country, though it cannot be generally adopted on all the lands of a farm, unless the whole is kept under a regular course of tillage and pasture. The apprehension, in England, of tenants over-cropping, has necessarily occasioned their being prohibited from breaking up meadows and pasture lands, whence, at this time, and at all times, a very considerable addition to the present supply of food might be obtained. The convertible husbandry, or the management of farms in an alternate course of

tillage

tillage and pasture, is in general well understood in the
highly cultivated parts of Scotland, from which system
the crops of artificial grasses are infinitely more abundant than from any mode of cultivation in England.

The articles most generally used as top-dressings are,
lime mixed with rich black mould—lime mixed with peat
—peat ashes—coal ashes—and soot.  The refuse articles
in different branches of manufacture, when they are to
be procured, are also attended with very beneficial effects; but as the quantity that can be obtained is very
small, they cannot be regarded as substances to be generally depended upon, though to lands in the vicinity of
such manufactories, they may with great advantage be
applied.

The ashes procured from peat, in the neighbourhood
of Reading in Berkshire, seem to possess a fertilizing
power, infinitely greater than ashes obtained from most
other peat. They certainly contain no alkaline salts; and
in an hasty analysis made about nine years since, no saline matter is recollected to have been got from them,
but a small proportion of Epsom salt.  Had these ashes,
however,

however, been analysed with more care, and when newly made, they probably would have been found to contain a hepar of lime, a salt which is soluble in water; whilst gypsum, to which it reverts on exposure to air, is insoluble. To this hepar may the fertilizing power of these ashes most probably be ascribed.

As mineral springs are frequently found to arise in peat mosses, it necessarily follows, that the ashes of different peat will contain very different saline and other matters. When a too large proportion of vitriol of iron, or green vitriol, is contained in peat, its ashes must of course be inimical to vegetation; but the injurious effects of this salt are to be corrected by the addition of lime, magnesia, alkaline salts, or dung. Of these substances, the preference is to be given to magnesia and alkaline salts; for, whilst they decompose the metallic salt, they form Epsom salt, Glauber salt, or vitriolated tartar, all of which are conducive to vegetation. The effect of dung on such ashes requires to be explained in a different manner. The iron, in this case, is by the application of the dung changed into a metallic state; whilst the vitriolic acid combines with the volatile alkali of the dung, and forms vitriolic ammoniac ; or by combining with the calcareous matter, and by the further aid of the inflammable or putrescent matter of the dung, it is changed into an hepar.

The burning of peat, for the purpose of procuring its ashes, must undoubtedly appear a very wasteful and dissipating process, when it is considered that there is seldom 1-20th of its weight in ashes procured by combustion. This process throws into the air the remaining 19-20ths of peat, which might, by other preparations of it, .be made to contribute, in a superior degree, to the purposes of vegetation.

The best preparation of peat, when intended for a top-dressing manure, is by the addition of alkaline salts or alkaline hepars, or by a mixture of glauber salt and hot lime with peat. As the soil cannot be injured by the application of alkaline salts, when mixed with a due proportion of peat, the quantity made use of will, in a great measure, be governed by the facility of procuring it, and the price of the article, and may, in some measure, be regulated by the quantity of alkaline salts capable of

being

being procured from the combustion of any crop produced from an acre of ground. In no instance the quantity of alkaline salts obtained will be found to exceed
one and a half cwt. unless from cabbages, turnips, and
potatoes. No restriction nor nicety need, however, be
attended to in the use of alkaline salts, except such as have
reference to cost, and the comparative beneficial results
from the increased produce of the ground.

The scarcity and high price of the above saline matters, in addition to a due want of knowledge of the proper
state which peat should be in, when these substances
are intended to be mixed together, have hitherto pre-
cluded the use of alkaline salts for the improvement of
the soil. The only substance of a caustic nature, and
capable of destroying the organic texture of vegetable
bodies, which in some places has been used, is lime,
though not in so judicious a manner as to insure at all
times, uniformly, good effects from its application, which
can never be depended upon, if the proportions of each
substance, and the particular state of them, are unattend-
ed to and neglected, as will appear by the following observations.

When

When hot, or newly calcined lime is broken into pieces of a small size, and mixed with peat, moderately humid, heat is disengaged, and that heat, by the slacking of the lime, when it is applied in too great a proportion, is so increased, as completely to reduce the peat to charcoal, and to dissipate, in a gassious state, all its component parts, excepting the ashes, part of the carbonaceous matter, and such a portion of fixable air, generated in the process, as is absorbed by the lime, by which that substance is made to return to the state of chalk. No benefit can, therefore, arise by this method of preparing peat with lime, the object not being to destroy and dissipate in a gassious state the component parts of the peat, but to make such a combination with the lime, and the gas generated in the process, as will, on the application of the mixture to ground, promote the growth of plants.

This object is best attained by mixing newly made, and completely slacked lime, with about five or six times its weight of peat, which should be moderately humid, and not in too dry a state. In this case, the heat generated will be moderate, and never sufficient to convert the

peat

peat into carbonaceous matter, or to throw off, in the state of fixable air, the acids therein contained. The gasses thus generated will be inflammable, and phlogisticated air forming volatile alkali, which will combine, as it is formed, with the oxygenated part of the peat that remains unacted upon by the lime applied for this especial purpose, in a small proportion. By this mode of conducting the process, a soluble saline matter will be procured, consisting of phosphat and oxalat of ammoniac, whose beneficial effects on vegetation have already been described.

Inattention, or ignorance of these important facts, has probably, in many cases, defeated the wishes of the farmer in the application of this preparation, which is particularly recommended as a top-dressing to grounds under pasture. The proportion of the lime to the peat here given, should be carefully attended to, and the mixing of the two substances together should be performed under cover, in a shed or out-house constructed for that purpose, as too much rain, or a too great exposure to air, will prevent a due action of the lime upon the peat.

The

The success of most operations, but more especially of those of a chemical nature, greatly depends upon a regular and due observance of circumstances apparently trivial.

This preparation of lime and peat is in a peculiar manner conducive to the growth of clover, and of the short, and as they are called, sweet kinds of pasture grasses. The soil also, by the application of it, acquires such a predisposing tendency to promote the growth of such grasses, as to prevent its growing afterwards rank, coarse, or sour herbage.

Notwithstanding that this preparation of lime and peat is certainly, when properly made, a valuable manure, yet the advantages that may be derived, by using alkaline salts instead of lime, are of much greater importance and general utility; in as much as the peat, by alkaline salts, is rendered completely soluble; whilst, by the application of lime, no greater proportion of it is made capable of solution than what is equivalent to the quantity of volatile alkali, which may be generated in the process; besides which, a large proportion of the acids con-

contained in the vegetable matter, combine with that which is calcareous, and form insoluble compounds.— What has been, and will be further stated on the different uses of peat, will, it is to be hoped, cause this valuable, and hitherto neglected article, to have a primary place in the farmer's calender, and teach him that nothing is useless, if man knew how to turn it to account. Many of the counties in England contain an abundant supply of peat, and it is still more generally, and in greater quantities, to be met with in Scotland and in Ireland.

In some few parts of Scotland, peat, in a natural state, or mixed with too small a proportion of dung, is applied as a manure. The growth of sorel is the never failing consequence of such application to soils, which contain little or no calcareous matter; whence may be inferred, that the soreline acid produced in consequence of the oxygenation of the peat, gives a predisposing tendency to the ground, on which it is applied, to promote the growth of sorel. Similar effects, in the production of sorel, have been observed to attend fossile coal, particularly at collieries where small coal has for a time been exposed to the action of air; hence also may be inferred, that

coal

coal consists of matters analogous to those of peat, and contains the basis requisite for forming the soreline acid.

The aptitude to receive and embrace useful and important ideas from small notices of things, must be acknowledged a faculty which is but seldom bestowed. In modern strictness of chemical theory and reasoning, this circuitous operation of nature, in promoting the growth of sorel, may probably be questioned; but as the diligent and minute attention of any individual, to any particular process which may have escaped the notice of others, ought rather to induce communication, than create an apprehension of hazarding the results of such inquiries, the above observation, like many others in this Treatise, is presented with no other view than that of calling attention to the minute circumstances daily presented by nature, and which are too frequently overlooked.

The promoting the more complete solution of vegetable substances, is undoubtedly one of the principal objects of the present design; still there are other circumstances of a more mechanical nature, which demand attention: such

as

as the proper mixture, or addition to soils, of substances capable of retaining a due proportion of humidity, and affording to plants the necessary supply of moisture. This is to be effected by a due application of decayed vegetables, (or as called in this Treatise) inert vegetable matter. A superabundance of this substance will cause the soil to be too spungy and open, and, on the alternate change from frost to thaw, to spew or throw winter corn out of the ground, as well as to injure or destroy cabbage, green kail, and other plants, produced by the alteration, which the water undergoes in such spungy soils, from a liquid to a solid state, and *vice versa*. Thus either chilling and rotting, or mechanically protruding or forcing the roots of the plants out of the ground. But to many stiff clayey soils, such vegetable matter may, in a peculiar manner, be serviceable, and is easily procured, in different states of preparation, from peat mosses.

The

The different preparations of peat which may be applied to ground, for correcting particular defects, or for rendering it generally more fertile, are,

Peat  -  unmixed

Peat  -  ditto    -    oxygenated

Peat  -  oxygenated, mixed with dung and urine

Peat  -  ditto      -      with alkaline salts

Peat  -  ditto      -      with alkaline hepar

Peat  -  ditto      -      with Glauber salt and lime

Peat  -  mixed      -      with newly slacked lime

Peat  -  and lime mixed with peat ashes.

All these preparations may, according to circumstances, be used to advantage. The consuming by fire peat to ashes, is always to be considered as the least productive, and most unœconomical. The most beneficial and productive of these preparations will be found to be,

Peat with dung and urine
Peat with alkaline salts
Peat with alkaline hepar
Peat with Glauber salt and lime
And peat with lime.

When

When the soil does not contain a due proportion of calcareous matter, a preference should always be given either to the last, or to the two last of the above preparations, until it shall have received a sufficient supply of an article so * indispensably necessary as calcareous matter to the production of sweet herbage, liguminous plants, and grain.

By the method heretofore recommended, of mixing peat with lime, in the proportion of five or six times the weight of the peat to the lime, no injury can arise by an overabundant use of this last-mentioned article. First, because the lime, when thus mixed with peat, meets proper matter to combine with, or to act upon; and secondly, the small proportion necessary to be used, when compared to the weight of the peat, will be considerably less than the quantity of lime only, usually given at one time to ground. The bulk, weight, and the expence of the preparation, will direct the farmer in the quantity requisite to answer his intended purpose. An application of a moderate quantity of lime, from time to time, whether mixed or not with peat, is much to be preferred

---

* To soils containing a superabundance, or a sufficient quantity of vegetable matter, a preference is to be given to lime unmixed.

ferred to the generally prevailing practice of laying lime at once, and in great abundance, upon ground, for these reasons :

First, The purchase is attended with a considerable advance of capital, and with great expence for labour and carriage.

Secondly, From the distance of the kilns, or places whence it is fetched, and the time it commonly lies on the ground to *slack*, it for the most part becomes *effette*, or in a great measure returns to its original state of chalk, before an opportunity offers for its being spread. And

Lastly, When easily procured, and properly *slacked* with water, immediately spread on the ground, and ploughed in, if applied in great quantities it will occasion a too immediate dissipation, in a gassious state, of the vegetable matters contained in the soil, from which the succeeding crop can only be benefited by the proportion it is able to receive during the dissipating process. Hence it is manifest, that an œconomical and frequent application of lime, in moderate quantities, either mixed with peat or other vegetable matter, or even by itself, is greatly to be be preferred to those abundant dressings of lime usually given at one time, which cause an action on the soil more powerful and violent than is conducive to, or compatible with a continued state of fertility.

In short, lime should be considered in a chemical and medicinal point of view, when so applied, acting as an alterative, corrector, and a decompounder; a disengager of certain parts of the animal and vegetable substances contained in soils, and as a retainer and a combiner with others; and is not to be regarded by the practical farmer as a substance fit for the immediate food and nourishment of vegetables, like dung, or decayed vegetable or animal matters. For, although calcareous matter, or lime, forms a component part of vegetable and animal bodies, still the quantity that can be obtained from the annual produce of most crops, from an acre of ground, will not exceed eighty pounds weight. This fact has been well ascertained, and if proper attention be paid to it, in regulating the conduct of the agriculturist, in the future application of lime, it will prove more satisfactory than all the chemical reasonings adduced in this Treatise.

By

By an over-use of lime in many parts of Great Britain,. the soil at this day requires a mode of treatment directly opposite to that practice. Some substance or substances ought to be applied to the soil, which would change the calcareous matter therein from its present state, or separate it from the insoluble combinations it has formed with the vegetable substances contained in the soil. This valuable improvement may be accomplished by a due application of acids, alkalis, and certain neutral salts, as before directed.

The effects produced on organic bodies by lime, clearly point out, that lime should never be mixed with dung, or with any substances which of themselves, or by the application of saline matters, would easily become putrid and rotten. Lime not only puts a stop to the putrefactive process, but disengages and throws off, in a gassious state, a certain portion of the component parts of such substances; whilst, with those which remain, it forms insoluble compounds that are incapable of promoting vegetation, until they are again decomposed and brought into action by other substances. In making of composts, rich surface mould is, of all substances, that which is most

proper

proper, when mixed in moderate quantities, to promote the dissolution and complete putrefaction of the dung. This process would be greatly accelerated by the further addition of a due proportion of the vitriolic neutral salts.

Methods of manufacturing, at a cheap rate, the most efficacious of these salts, have been discovered, and farmers will soon be enabled to make the necessary experiments, and to satisfy themselves that such salts possess the powers ascribed to them. The price will be regulated by the duty that may be charged on sea salt, and on the bitter refuse liquor of the salt works, whence these articles are to be made.

But as it may happen, that much time may elapse before any relief is granted, or any alteration made in the present salt laws, it becomes important to consider how a supply of salt, or what is still more valuable, how a supply of sea or salt water may be obtained for the use of cattle, and the purposes of agriculture, without being subjected to the present duties. Previous to this expla-

Q                                       nation,

nation, it is proper to state some further circumstances, more fully to impress on the mind of the farmer the effect which salt or sea water has in promoting the more full putrefaction of dung and vegetable matters.

It is well known, that ships built of unseasoned timbers are at first very unhealthy. The exhalation of the vegetable juices of the fresh wood is not the sole cause. It is principally to be ascribed to the putrescent hepatic gas, generated by the mixture of the vegetable juices with the vitriolic neutral salts contained in sea water, forming what is called bilge water. The smell of it is no less offensive, than its effects are prejudicial to the health of the ship's company. When a new vessel happens to be tight, and to make little water, it is the practice with all intelligent seamen, to sweeten the vessel's hold and timbers, by daily letting in and pumping out a sufficient quantity of water.

Certain gasses, which are injurious to the health of animals, are favourable to the growth of plants: hepatic air is one of them; and as hepatic air is formed in vessels

holds

holds by the action of sea water on the soluble matter of the wood, the same effect will be produced by the addition of salt water to dung or to vegetable matters. The generation of the hepar is to be ascribed solely to the vitriolic salts contained in sea water, and there is some reason to suspect, that sea salt, or muriat of soda, may suffer a decomposition in this putrefactive process, and that the marine acid thereof may be decomposed.

The putrefaction of sea water is not confined to the bilge water in vessels. The water of the sea itself, in certain southern latitudes, undergoes a material change, emitting, during long calms, a putrid offensive smell; and water intended for the purpose of making salt, kept too long in the reservoirs during summer, will suffer such an alteration in its nature, as to be rendered incapable of yielding chrystals of sea salt. A month or six weeks of warm weather is in this latitude sufficient to produce the change, which is prevented by letting out of the reservoir, every fourteen days, part of the old brine, and taking in a fresh supply of sea water, frequently very inferior in concentration or strength to that which is

obliged

obliged thus to be discharged. If this tendency to putrefaction take place simply in consequence of the small
proportion of animal and vegetable matters contained in
sea water, there is still greater reason (exclusive of actual experiments) to conclude, that it will take place, in a
much higher degree, on adding sea water to a larger proportion of such substances as of themselves have a tendency to the putrefactive state. As any further proofs of
the effects of the saline matters contained in sea water,
in promoting putrefaction, may be deemed unnecessary,
a method of procuring a supply, without incurring the
expence of manufacturing them, or being liable to the
present duties, is an object of the greatest importance to
the farmer and the grazier, particularly to those who are
at a distance from the sea..

In its vicinity, farmers and others may avail themselves
of their situation, and procure sea water either to mix
with dung, or for the other purposes to which the
application of it has been recommended. A ton of
sea water contains from a bushel to a bushel and a quarter of sea salt, beside a certain proportion of the vitriolic
salts..

salts. This quantity could not be purchased in England, including the duty, at an expence less than seven shillings, which farmers, situated as before described, may procure at the small expence of carriage.

Sea water may be raised, where coal is cheap, by means of a fire engine, to such a height as, corresponding with the level of the inland country, would allow the water to be conveyed in small open canals, in wooden or in earthen pipes, to a considerable inland distance; each farmer, or proprietor, receiving as it passes the necessary supply.

A fire engine of forty inches cylinder, with ten inch pumps, would raise to the height of 30 fathoms, 1000 tons of water per day, equal to 6000 tons per week, or 300,000 per annum. Admitting this water to contain 1 in 33 of salt, 9,636 tons would annually be procured; and supposing each farmer, on an average, to receive such a quantity of sea water as would be equivalent to 33 tons of salt, 9,636 tons would be sufficient to supply 292 farms. Should the interest of the capital employed, and the

the charges attendant on the fire engine amount to 500l. a year, which is an abundant allowance where coal is cheap, and should the further interest on the sum engaged in conveying the water to distant places amount to 500l. more, the total expence for raising, conveying, and serving 292 farms with sea water, would amount to 1000l. per annum, or something less than 3l. 10s. for 33 tons of salt contained in the sea water which each farm would receive. The prime cost of sea salt, exclusive of duty, could no where be less than 1l. in some places 3l. on an average it should not be reckoned at less than 1l. 10s. per ton, which on 33 tons would amount to 49l. 10s. But, as to this must be added the present duty of 10l. per ton, it would cost the farmer 379l. 10s. for the quantity which, by these means, he may receive for about 3l. 10s. Admitting, however, that the expence of the fire engine, and of conveying the water, were to be double the sum above stated, and that instead of 3l. 10s. it were to amount to 7l. a quantity of salt water containing one ton of salt would, in this latter case, cost the farmer only 4s. 3d. or at the rate of one penny farthing per bushel of salt of 56lbs. whilst the same quantity would now cost 6s. Farmers,

who

who are situated near the sea, should lose no time in satisfying themselves of the beneficial effects that may arise by a proper use and application of salt water to dung heaps, compost mixens, and to other purposes: it would be highly praise worthy, were some intelligent persons, after carefully making such trials, to publish the result of their experiments for the information of others.

Should the advantages attendant on the application of salt water, induce its being generally made use of near the sea coast, there can be little doubt that cheap methods would soon be devised, for conveying from many places salt, or sea water, to very considerable distances.

Every plan or method whereby a greater quantity of dung may be made, should be carefully attended to. A considerable increase in the quantity of that article, as well as an œconomical saving, and benefit in other respects, will arise by feeding the working horses, and perhaps part of the neat cattle of a farm with
clover,

clover, tares, saint foin, and other green crops, in the stables, in out-houses, or in the farm-yard, instead of turning them out to pasture, as is the general practice. By this œconomical and judicious method, no part of the crop is spoiled on the ground, but every part thereof is made to serve the intended purpose. A much larger quantity of dung will thus be made, of a richer and superior quality, in consequence of the eggs deposited by the various flies, and the maggots and worms bred therein during the summer season, than can be procured in the winter time. By a flooring of clay or chalk under the pavement of the stables, out-houses, &c. the urine of the cattle will be prevented from sinking through, and by the same precaution the valuable juices of the dung heap may also be prevented from soaking into the soil of the farm-yard, especially if due care be taken to add from time to time a sufficient quantity of peat, or rich black mould, to absorb or suck up the surplus moisture produced by the succulent food. Thus may much expence be spared, a larger proportion of dung be acquired, and a more plentiful supply of food be provided for cattle, without any fatigue or

loss

loss of time in roaming about to procure it for themselves.

Experience only can teach, or warrant the belief of
how few acres of ground, under the culture of artificial
grasses, when cut green, and daily given to working horses
and other cattle, will suffice for their maintenance. The
artificial grasses, or plants, best adapted to this purpose,
are, red clover, tares, and saint foin. None of these succulent plants with large stems and leaves, answer so well
to be depastured as to be mown : not only on account of
the injury they receive in being bruised by the treading
of cattle, but by being constantly cropped and kept short,
they are deprived of the nourishment which they principally receive by their stems and leaves. Saint foin is
best suited to chalky or dry soils, and to the southern
parts of Britain. It has often been tried without success
in the northern parts of England and in Scotland.
Winter tares have also been sown, but have not been
found to answer any valuable purpose. Clover and summer tares, therefore, should be the only plants of which
the cultivation on a large scale should in these parts be
attempted, and every prudent farmer will take care to

have

have a full supply of them, as in the event of a superabundant quantity for green food, these crops are equally
proper for hay. Tares should always accompany the
culture of clover, to supply the deficiency of herbage
between the first and second cuttings of the clover. When
black mould has been worn out by repeated tillage, and
is filled with root and seed weeds, no preparation or
mode of culture can be adopted that will more effectually
clean and fit it for a subsequent crop than tares, which
should be sown thick, that a luxuriant crop, completely covering the ground, may be insured.

It is a matter of great importance (though seldom
attended to) that the food intended for working horses
should be so prepared, or of such a nature, as to allow
them quickly to satisfy their hunger, that more hours
may be allotted for rest during the interval afforded
from labour. When thus fed in stables or farm-yards
with green clover, tares, saint foin, &c. unless on extraordinary occasions, they do not require oats or other
grain; but in the winter, or other seasons, when fed on
dry hay or straw, and when occupied in the business of
the farm, a supply of oats, other grain, or nourishing food

is

is found to be indispensably necessary. Oats mixed with beans or pease, is the grain generally given, although, when barley is at a certain price, it would be an object of œconomy to make use of it in lieu of oats. The horses in Spain and Portugal are exclusively fed with barley, to which they give a preference.

To shew that this taste is not peculiar to foreign horses, it is a well known fact, that when Burgoyne's regiment of light dragoons went to Portugal in the year 1761, the oats sent from England were carefully reserved for the opening of the campaign, and in the mean time the horses were supplied with barley, of which they became so fond, that when the campaign began, they rejected the English oats, and were afterwards uniformly fed with barley; notwithstanding which, it will be no easy matter to eradicate the prejudice in favour of oats from the minds of English grooms, which can only be effected by actual trials and experiments, made by the intelligent, who will afterwards decide for themselves. The different proportions of meal, or farinaceous matter contained in barley and in oats, ought not to pass unnoticed.

A Scotch

A Scotch boll of good barley, (equal to six Winchester bushels) will weigh eighteen stones Dutch weight, the proportion of husk to the kernel or meal does not exceed a stone and an half, or two stones; whilst the like quantity of oats, weighing fourteen stones, will not produce more than eight stone of meal or kernel. As the average prices of barley and oats may be fairly stated at seventeen shillings and twelve shillings per boll, it is plain, that in the one case sixteen stones of barley meal are purchased for seventeen shillings, whilst eight stones of oat-meal will cost twelve.

Experience has proved, beyond a doubt, that all corn given to horses and other cattle, should be broken, by being passed through rollers, or crushed in a mill. Horses, not being ruminating animals, will receive considerable benefit by this œconomical operation, and the loss or waste, by swallowing their corn whole, will be prevented. Barley boiled in sea water, or with a due proportion of sea salt, is a good supper for hard working horses. These œconomical modes of feeding horses have been confirmed by several years experience, in keeping the horses of a colliery.

It

It is not, however, to be understood, that a preference should be given to the feeding of horses with broken or ground corn, instead of potatoes or carrots, which judicious application of these roots is now becoming very general. Nothing farther by the above statement being meant, than, that those who may prefer the use of grain, should be made acquainted with the best and most œconomical manner of giving it: and in the same point of view is to be regarded the following observations on feeding horses and other cattle on malted corn.

There is great reason to believe that the most judicious method of feeding horses or cattle with corn, is by giving them malted instead of raw grain.——— Malted corn tends to open the body, and cleanse the intestines from all putrid saline biliary obstructions; which effects being attained, it no longer operates in this manner, at least in no degree inconsistent with the health of the animal. To such as do not regard a small expence in ascertaining so valuable a fact, it is earnestly recommended that a comparative trial be made in feeding two teams of horses with malted and unmalted grain, of the same sort and quality. Three months would fully ·

as-

ascertain this question, which, in the event, proving as it is here presumed it would, highly in favour of the malted food, there can be little doubt, that on proper representation of the beneficial effects of malted corn in feeding cattle, Government would permit its use for this purpose, under certain restrictions and regulations, especially as the indulgence is capable of being rendered an object of considerable revenue, as will appear in the miscellaneous observations of this work.

The ramifications, which, in a discussion of this nature, branch out and involve almost every circumstance that respects the œconomy of rural affairs, render it no easy task to preserve such a distinct arrangement of the several subjects as may be generally approved.

The feeding of cattle may, perhaps, be considered as improperly introduced under the last title of " Stable, Farm-Yard Dung,* and Composts :" but on reference to the

---

* The practice of haulming, or mowing stubbles, is in most cases to be recommended ; as the farmer, by laying the stubble so gathered in his farm-yard, will be enabled to return a greater quantity of prepared manure to his fields, than if he had ploughed the stubble in.

the reply of CATO, assumed as a text in the beginning of the chapter, it will be obvious, that the judicious Roman, by his expression of " *bene pascere*," (which is to be translated, to graze well, or to procure food for cattle) must have had in view the connection between feeding of cattle and the production of manure. Every article of manure, ultimately tending to render ground more fertile and productive, falls likewise under the present discussion.

Insects of all kinds, under the names of worms, snails, grubs, slugs, &c. &c. exist in the ground ; and in such grounds as are fertile, in much greater abundance than is generally imagined. Their food, most undoubtedly, must be either fresh vegetables, or decaying or decayed vegetable matter. In the former case, such insects prove extremely destructive ; whilst in the latter they may be of service to the vegetable kingdom, by rendering the decayed or decaying vegetables, eaten by them, more soluble by the process of digestion. In this class are to be included the common earth worms, which are only to be found in great numbers in ground containing

a large

a large proportion of vegetable or animal matters. Worms of this kind feed only on rich earths; and as they are never found on sterile ground, their nourishment must necessarily depend on the before-mentioned subtances contained in the soil.

The excrements of these worms appear on the surface in great abundance; particularly in moist weather, succeeding to a long drought; or at the season of the year when the dews fall heavily. On these occasions the worms rise to the surface, for the purposes of engendering, supplying themselves with moisture, and of voiding their excrements. These excrements, from the astonishing number of worms contained in rich ground, cannot but promote vegetation; though a temporary inconvenience may sometimes be incurred, by preventing the cattle from freely depasturing, when the surface is too much covered therewith.

All insects or worms in the ground, as well those which apparently are of no disservice, as those that are known to be noxious and destructive to the roots, stems, and

and leaves of vegetables, may be destroyed by alkaline salts and hot lime; which substances have the power of dissolving the continuity or texture of organic bodies, and are particularly fatal to the soft bodies of living insects. Insects are likewise to be destroyed by neutral salts, and by saline bituminous substances. The bodies of these insects, when dissolved by putrefaction, become, like other animal matters, serviceable to vegetation.

The vitriolic acid will also act in destroying insects and other animal substances, in a manner somewhat similar to alkaline salts, with this difference only, that the one forms an acid, the other an alkaline sapo.

Vitriolic acid, diluted with a due proportion of water, and superacidulated vitriolic salts, may likewise be used with a double effect, in the destruction of insects, in ground long under cultivation, and which contains much animal and vegetable matter, in the state of phosphat and oxalat of lime. In this case, not only the insects will be killed, but the vitriolic acid will, by superior affinity, combine with the calcareous matter of the phosphat and oxalat of lime, whose disengaged acids will form new

soluble,

soluble, fertilizing saline combinations with the ammoniac or volatile alkali and magnesia that may be contained
in the soil.

Sea salt is found to destroy snails, slugs, grubs, worms,
&c. by making them void the contents of their bodies,
evacuations too powerful for them to withstand. By
these means, not only their bodies, but their evacuations soon become food for vegetables. It is principally
in this point of view that any benefit can be attributed
to the action of the muriat of soda, or *pure* sea salt,
upon ground.

Soot is used in many parts of Britain, with very beneficial effects, for the destruction of the wire worm, and
other insects, which prey upon the young tender leaves
and roots of plants. This article, exclusively of carbonaceous and earthy matters, consists of mineral or resinous oil, rendered capable of solution in water by
the saline matters, which are contained in soot in
great abundance. The solution is an extremely bitter
high coloured oily liquor, which not only poisons the
insect on which it may fall, but also communicates a
bitter taste to the surface of the roots and leaves of
plants;

plants; thus rendering them unfit for the food of such insects. The effects of soot are not confined to the destruction of insects alone; for the resinous bituminous oil, dissolved by the saline matters, promotes the growth of plants in a very high degree, and by rapidly pushing them to an advanced state of maturity, allows no time for the ravages of insects on the young and tender roots and leaves of plants, on which, in general, they principally feed.

The quantity of soot that can be collected, is so inconsiderable, in proportion to its uses, and the demand there might be for it, were its powers more generally known, that a method of procuring a greater supply of this essential article, or of any substance that would produce similar good effects, is an object of very great importance to Agriculture. This is to be accomplished by different preparations, amongst which coal-tar is to be included. The chemical analysis, and deleterious or poisonous effects of coal-tar on insects, prove its identity to soot; the only difference being, that of its containing less fuliginous, or carbonaceous and earthy matters : in which respects, as carbonaceous and earthy matters are not soluble

in

in water, coal-tar,* weight for weight, is to be preferred to soot, for the destruction of insects, or as an article of manure. But as the quantity of coal-tar that is now capable of being made in Britain would be insufficient for the purposes of Agriculture, it has led to the discovery of a cheap method of preparing *other substances*, which, most probably, will be found equally, if not more efficacious, as manures, and in the destruction of insects.

These preparations may be sown by hand on the ground, and there is every reason to believe they will succeed, not only in the destruction of the wire-worm, so, injurious to the roots of grain, and to the fly which preys upon the tender leaves of young turnips, and other vegetables, in this country; but likewise to the ants and sugar-cane borer, in the West Indies, and the Hessian fly in America.

* Although the good effects of coal-tar, in repeated instances, have been manifested, yet it is now entirely out of use for vessels bottoms or sheathing, on account of the *protection* it affords them from the attacks of the teredo or gribble worm, for at least thrice the time which vegetable tar does. This has been the reason assigned for its disuse by some of the most considerable and *candid* ship-builders and repairers of ships in the River Thames, and other places in England. Vessels have been known to perform six voyages to the West Indies with the same sheathing, when payed with coal-tar.

DRAIN-

## DRAINING.

Draining of ground in the northern and humid climate of Britain and Ireland is indispensably necessary, and is the precursor to all culture and improvement of the soil.

The principal oject in draining, is to free the surface from moisture at those seasons when it would prove hurtful.

Water constitutes the chief food of plants : it is decomposed in the process of vegetation, the plant retaining the hydrogen or inflammable air, as well as the calcareous matter held in solution in the water, whilst the oxygen or vital air, the other component part of water, is disengaged.

Certain degrees of cold prevent the absorption of water by vegetables. Water, during the continuance of such degrees of cold, is of no service; its presence at
those

those times generally proves in inimical to the future growth of plants. On this account, rain, during the cold and chilling winter months, is injurious; whilst warm summer showers are attended with a contrary effect: hence, when vegetation is not advancing, or but slowly proceeding, the ground should be kept as dry as possible.

Considerable benefits ensue to certain soils by artificially watering the ground at certain seasons, yet much greater, and more extensive advantages to the soil, and to the agriculture of these kingdoms, would result by a more complete and general drainage of the surface. What is most important to be done, should ever be done first, and the attention of the farmer should not be called away to other objects, such as the irrigation of meadows, until he had relieved his lands of the injurious surface water, and laid them sufficiently dry; previous to which, the full benefit that may arise by irrigation, or the judicious application of water at certain seasons, cannot be expected.

The different mechanical operations of draining land are in general so well understood, that it cannot be deemed

necessary

necessary to enter into a description of them, in a Treatise which has principally in view such circumstances and matters as have not hitherto, on chemical principles, received satisfactory explanation. Under-draining at this time is very deservedly in great estimation, and is becoming the general practice. It saves much waste of ground, more completely answers the intended purpose, is of longer duration, and ultimately less expensive than any other kind of draining. A method has lately been discovered, and practised with success, by which, in many places, the upper stratum is drained by the assistance of the mineral strata beneath it, through which the water is made to drop, and in this manner taken from the surface. Draining is not only to be accomplished by these judicious methods, and by open drains, but lands are to be made dry by such a mechanical alteration in the component parts of the soil as render it less retentive of moisture. In stiff lands this may be effected by lime, chalk, marl, coal-ashes, brick-dust, or calcined clay, and by sand, when applied in *large* quantities, whilst the too great tendency in sandy or light soils to part with moisture is to be corrected by other applications.

FAL-

## FALLOWING.

It has been frequently noticed in the preceding pages, that alkaline salts act more powerfully on some kinds of peat and inert vegetable matters than on others, particularly on those which become oxygenated by being exposed to the action of air. This points out, that the practice of fallowing ground containing much vegetable matter, by repeatedly exposing fresh surfaces to the action of the air, occasions the peat, or vegetable matter, to be more easily dissolved, or acted upon by alkaline salts; but when no such application is made, the insolubility of the vegetable matter is by fallowing increased, which, to certain grounds, may prove, instead of a benefit, a real injury.

The putrefaction or solution of vegetable substances is more readily promoted by a close or stagnated state of the air, than by a constant supply and addition of oxygen or pure air, as happens to vegetable substances when subjected to the operation of fallowing.

Clover,

Clover, saint foin, cabbages, turnips, leguminous crops, hemp, and those plants which overshadow the ground, and cause a stagnation of air, prevent thereby the excessive exhalation of moisture, and promote the putrefaction or decomposition of vegetable matters contained in the soil: such crops will therefore prove more œconomical and beneficial to subsequent crops, than the present system of fallowing.

By fallowing, not only one year's rent and labour are lost, but likewise the vegetable matter contained in the soil is thereby rendered less fit to promote the growth of subsequent crops. Fallowing should be practised but sparingly; its principal use is in altering the mechanical arrangement of the soil, either by pulverising it, or making it more compact, (both of which effects, according to circumstances, are thereby produced) and in destroying root, seed weeds, or insects. These objects being attained, recourse should never be had to the same operation, unless it becomes requisite from the failure of crops, or other incidental causes, which are best provided against by substituting the culture of drill crops instead of a fallow.

T                                        To

To soils which contain much inert vegetable matter, it is probable that advantages would be derived from umbrageous green crops without fallowing, equal to those experienced, when hemp is made to precede a crop of wheat; without which preparation this crop would not have answered the expectation of the farmer.

It is therefore obvious, if ground receives benefit by being overshadowed, the same ground must receive injury by a direct contrary mode of treatment.

PARING

## PARING AND BURNING

Is a comburatory dissipating process, whereby nine-
teen parts out of twenty of the vegetable matter, the
only substance the fire can act upon, is dissipated and
thrown into the air.

This process, no doubt, from its having been carried
to excess, and so often repeated as to destroy all the vegetable matter contained in the soil, has, it is said, been
prohibited by the Legislature of Ireland, under a penalty
of ten pounds per acre.

Moors overgrown with ling or heath, peaty soils, or
soils covered with a sward of coarse unprofitable herbage, and containing a superabundance of vegetable
matter, may, with due precaution, be subjected to this
process with very beneficial effects.  It may likewise be
attended with advantages to strong clayey soils, from the
effect that burned, or half burned clay has in rendering
such soils more open and less tenacious; in which case

T 2                                the

the benefit arising from the mechanical arrangement of the soil will probably more than compensate for the dissipation of the vegetable matter of the sward. It would, however, be more œconomical, when the soil is thus intended to be made more open, to calcine the clay in clamps or kilns, and to spread it afterwards on the ground, either by itself or mixed with lime.

Paring and burning is the process, which, in the cultivation of peat mosses and fens, is made to succeed that of draining. Care should be had to burn only as much of the peat as will yield the proportion of ashes necessary to alter the arrangement of the parts of the soil: an effect which, with still more advantageous consequences, is to be attained by lime, lime-stone gravel, or even by common mould or soil.

It is only from the ashes of fresh or growing vegetables, that saline substances or alkaline salts are to be obtained; none can be got from peat or decayed vegetable matter. The proportion of alkaline, or other salts, produced by paring and burning, is so very small, that were the benefits immediately resulting from paring and burning

burning to be ascribed solely to these salts, the purchase of them at the market price might, perhaps, be more œconomical.

The saline matter produced in the process of paring and burning, for the most part consists of vitriolated tartar—the alkali of the burnt vegetables, combining with the vitriolic acid, which in different states of combination is contained in most soils. Vitriolated tartar has very powerful effects in promoting vegetation; but as it is not to be procured in sufficient quantities to answer the purposes of agriculture, the deficiency is to be supplied by Epsom and Glauber salts, which produce effects equally beneficial when applied to ground.

Although paring and burning has by many persons been much recommended, still it requires great limitations or restrictions. In some cases it may be proper; while, in the hands of the unskilful, it may be attended with the most pernicious consequences.

If all the benefit that can be derived by this practice, may hereafter be attained by the application of lime, alkaline

kaline salts, neutral salts, &c. &c. without risk of any of the attendant disadvantages on the process of paring and burning, a decided preference of course will be due to methods that render this practice less necessary.

———  ——————

THE simple earths, air, water, saline bodies, vegetable substances, &c. &c. having thus been considered, as far as the properties of each relate to the present design, it is now become necessary, previously to any further discussion respecting the practical part, to give such directions to the cultivators of the soil, as may enable them to ascertain the nature and proportions which the component parts of it bear to each other; and consequently the value of the surface mould contained in the different parts of their farms or estates; and how, by this information, they may be enabled to apply with most advantage the several ameliorating substances herein recommended.

It has not been, nor would it be possible to avoid making use of chemical terms, consistently with the plan of a work

a work, which has for its object the making every farmer, to a certain extent, a chemist, so that he may be
enabled to understand the nature and properties of the
several substances, in the management of which he is
daily engaged ; and that in all his future attempts to improve the soil, the success of his operations may no
longer depend on guess-work, or on chance, but be
regulated by a proper knowledge of the materials he
may have to work with—how each may best be applied
or acted upon, and what effects will ensue from their
different combinations.

Cultivators of the soil should be able to distinguish,
by chemical tests, the proportion of the following substances in different soils, viz.

> Clay,
> Chalk,
> Sand,
> Magnesia,
> Earth of iron, and
> Vegetable matter.

They should understand the properties and effects, and superior affinities of alkalis and acids; as well as the names, properties, and compounded electrive attractions attendant on the mixture of the different neutral salts, and their effects on vegetation. They should be well acquainted with the powers of lime, and should clearly and distinctly comprehend the putrefactive and oxygenating processes; as well as the consequences resulting from the action of fire on the vegetable matter contained in the soil.

The first step that a cultivator of the ground should take, when possessed of the above information, is to ascertain by experiments, in what proportions chalk, clay, sand, magnesia, and vegetable matter exist in the soil, in the different parts of the farm he purposes to cultivate; in order that he may, from such information, be enabled to administer to each part those particular substances that it may require, to constitute it rich and fertile mould. A soil of this description ought to contain a due proportion of the simple earths, and of the remains of vegetable and animal bodies.— To enable him to make the requisite experiments, he

he should procure the following articles and vessels :

Two sets of small scales and weights, one to weigh a few
pounds at a time, and another smaller and more accurate,
for ounces and grains: some porcelaine glass, or stone-
ware vessels unglazed, such as are called Vauxhall ware :
some muriatic acid, and mineral alkaline salt. These being
provided, the method of proceeding to ascertain the different proportions of the different substances in soils, is
as follows :

The presence of calcareous mattter is ascertained, by
applying to the mould suspected to contain it, some
marine acid diluted with water. If it contain calcareous
matter, an effervescence will take place, and a neutral salt,
called muriat of lime, will be formed. This is to be separated from the earthy insoluble matter, by a due proportion of water, and is to be evaporated to a certain degree.
Lastly, the calcareous matter is to be precipitated by mild
mineral alkaline salt. When the calcareous matter thus
precipitated shall be collected, washed, dryed, and weighed,
the quantity contained in the soil will be ascertained by

U                                          the

the proportion it may bear to the weight of the *dry* mould on which the experiment had been made.

The same process and the same acid will serve to show if magnesia be contained, and the proportion it may bear to the soil. Magnesia is not in general found in any very great proportion in surface mould, although there is more of it contained in ground than is generally imagined. It will, for the most part, be found accompanied by calcareous matter; and as both these substances, when dissolved by the marine acid, are very soluble, and blended together, a separation is to be effected by the following process.

The earths of magnesia and calcareous matter are to be precipitated by mild mineral alkaline salt. The precipitate, or earthy residuum, when washed, is to be dissolved by a due proportion of the vitriolic acid diluted with water. With the calcareous matter it will form gypsum, (a very insoluble salt) whilst with the magnesia it will form Epsom salt, a salt of great solubility. These salts are to be separated by priority of chrystallization; and their respective weights being ascertained, when deprived

deprived of the water of chrystallization, and brought to an equal degree of dryness, the quantity of calcareous matter and magnesia in each may be ascertained by BERG-MAN's or KIRWAN's tables of the proportion of acid, alkali, earth, and water contained in different neutral salts. To those who are not provided with such tables, it may suffice to say, that

|  | Acid | Calcar. Matter | Water |
|---|---|---|---|
| 100 parts of gypsum contain | 48 | 34 | 18 |

|  | Acid | Magnesia | Water |
|---|---|---|---|
| 100 parts of Epsom salt contain | 33 | 19 | 48 |

As both clay and sand, in different proportions constituting either a clayey or sandy soil, are distinguishable by the sight and touch, there is no occasion for giving any chemical test, to prove their presence. The proportion of the coarser parts of siliceous matter or sand, in soils or mould, may be ascertained by washing.

The presence of vegetable or carbonaceous matter in surface mould, when in any considerable proportion, is apparent, either from its black colour, or from the vege-

U 2                                    table

table matter, appearing in the soil in an undecayed state. Chemical tests, in either of these cases, are unnecessary. When it may be requisite, however, to ascertain the presence or proportion of it in clayey or other soils, in which, from colour or extreme division of parts, it is less apparent, it is to be done in one or other of the following methods :

By properly drying and weighing a certain weight of mould, and then submitting it to such a degree of heat as will consume the vegetable or carbonaceous matter to ashes : at the same time, the heat must not be such as will disengage the fixable air from any calcareous matter ,or magnesia that may be contained in the mould or soil submitted to trial. The difference in weight between the dry mould, and that which is thus submitted to the action of fire, will be the proportion of vegetable or carbonaceous matter.

It is likewise to be done by melting some salt-petre in an iron laddle, bringing the salt-petre to a red fusion, and then dropping into it, by little and little at a time, the earthy matter, taking care previously to dry it tho-

thoroughly. The dropping in of the dried mould should be continued until the complete deflagration of the salt-petre is effected.

The practical observation to be deduced from the above experiment, is, that the soil or mould which contains the most vegetable or carbonaceous matter will deflagrate the greatest quantity of salt-petre; or, in other words, that it will require less mould to deflagrate a given weight of salt-petre, in proportion as that mould contains a greater proportion of inflammable matter.

The presence and proportion of vegetable and inflam-mable matters in clay may, in some degree, be proved and ascertained by the degree of blackness in the colour, which the interior parts of the clay assume, when sub-jected in the fire to a certain degree of heat.

The existence and proportion of most saline matters in soils are to be discovered by lixiviation, with warm water, and by subsequent chrystallization.

Gypsum is to be detected by boiling the earth with al-kaline salts; in which case, the gypsum will be decompos-ed, ed, and the vitriolic acid of the gypsum will join with the mineral alkali, forming Glauber salt, which is very soluble. The quantity of gypsum previously existing in the soil is to be ascertained by weighing, when properly dried, the calcareous matter which had been precipitated by the alkali; and by adding thereto, in calculation, the proportion of vitriolic acid necessary to constitute it gypsum; having previously deducted therefrom the proportion of fixable air which the precipitated chalk contains. The proportion of fixable air and vitriolic acid contained in chalk and in gypsum are in the proportions as here stated :

|  | *Fixable Air* | *Calcareous Matter* |
|---|---|---|
| In chalk, | 43 | 53 |
|  | *Vitriolic Acid* | *Calcareous Matter* |
| In gypsum, | 48 | 34 |

The following is given as an example of the method of making this calculation:

Residuum

*Grains*

Residuum of precipitated chalk,   -  -    480

Proportion contained therein of fixable air,   212

---

Calcareous matter,  -  -  -  -  -   268

Proportion of vitriolic acid necessary to constitute

gypsum with the calcareous matter,   -   -   354

---

Total quantity of gypsum,   -  -  -   622

SOIL.

## SOILS,

### *ARGILLACEOUS OR CLAYEY.*

THERE is no clayey soil that is pure and free from sand; and there are but few clays that are free from a mixture of calcareous matter, magnesia, vegetable and animal matters, mineral oil, and other mineral or metallic substances: some clays are of a much more unctuous, and, as it were, greasy nature, than others. They do not differ more in this respect, than they do in the appearance they assume when submitted to a moderate degree of heat. Those clays which are the most unctuous and greasy to the touch, are by calcination changed to a black colour. This must be owing either to their containing animal or vegetable matter, although previous to calcination it escapes observation; or the inflammable matter in the clay may exist in the state of a colourless mineral oil, adhering obstinately to the clay, and not capable of being separated from it by water, with which oil can hold no union; yet capable of being changed into a black carbonaceous matter by the action of fire. A due mixture of clay serves the important purposes of retaining

in

in the soil the attenuated vegetable and animal substances, as also the mineral oil. Of this description are those
clays, or clayey loams, which have been deposited by the
sea or muddy streams, containing a considerable proportion of the exuviæ, or remains of animal and vegetable
bodies, in an extreme degree of attenuation. Such
soils as these are the most permanently fertile, and
where the climate is favourable, produce the heaviest
and best filled grain. Soils formed by depositure, for the
most part contain a sufficient quantity of calcareous
matter. Adding lime to such lands may prove injurious,
by its expending, taking up, or otherwise altering the
arrangement and combination of the animal and vegetable matters, which should carefully be preserved for succeeding crops. Under any circumstances, lime should be
given to such soils but sparingly.

There are clayey soils containing little or no animal,
vegetable, or bituminous matter, and which are equally
deficient of calcareous matter, consisting only of clay, sand,
and the earth of iron. To improve and render fertile a soil
of this description, is truly an herculean task, and will
seldom repay the industry of the cultivator, unless situ-

ated

ated in the neighbourhood of a town, where more dung
may be procured than can be spared from the farm in its
contiguity. A soil of this nature can receive little or no
benefit by the application of lime, as it contains nothing
for the lime to act upon or combine with. When under
such circumstances, that dung, or such like manure, cannot be procured, a preparation of peat, with the moderate proportion of lime before directed, seems to be the
next best application. A soil of poor lean clay, such as
above described, will require eight tons of lime, and
forty-eight tons of peat, for one dressing. Doing things
partially can never answer: this quantity is the least that
ought to be applied; a much greater may be given, if the
articles can be cheaply and easily procured. In this the
farmer will be regulated, in a great measure, by his ability of doing, or extent of his capital. His primary object,
in this case, should be to promote the growth of pasture
grasses, because the soil at first will be in no heart to produce crops of grain; and, secondly, because the promoting
the growth of such grasses, and judiciously depasturing
and folding, is the surest way of improving such lands.
After the grass has taken hold of the ground, and is beginning to carry a tolerably thick sward, its thickness
and

and quality may be greatly improved by some one or more of the top-dressings or preparations before recommended.

There is a very great extent of poor clayey soil, similar to that which is here alluded to, in many parts of the North of England and in Scotland, for the most part lying at a considerable height above the level of the sea, and frequently in the vicinity of peat mosses, whence it might be supplied with vegetable matter. There are computed to be in the county of Lanark, or Clydsdale, 40,000 acres of peat moss totally unimpoved, producing nothing itself, nor contributing in any way to the fertility of the adjacent poor lands, which are as destitute of vegetable matter, as the moss contains a superabundance. Mr. Næsmith, the Agricultural Surveyor of that County, judiciously makes a remark on the injurious effects that such mosses may have on the climate of the adjacent country. It requires a much longer time, and a much greater application of dung and vegetable matters, than would be generally believed, before poor lands of this description can be rendered highly fertile, and made in all respects similar to land that had been long, or for ages under cultivation. Ten times the quantity of peat or vegetable matter recommended to be given at once, or 480 tons, would scarcely bring poor barren land to the colour of rich black mould, known in Scotland by the name of Infield land, and to which, for ages, the dung of the farm has been exclusively applied.

Experiments made with an intimate mixture of poor, lean clay and peat warrant this assertion; here purposely stated, that the over sanguine cultivator, or improver of ground, may not imagine, that with a summer fallow, and a dunging, or dressing or two, he may be enabled to complete so arduous a task. Land is always requiring a supply of manure, and repays in general more abundantly for the last expence, when brought to an advanced state of cultivation, than for that which at first is incurred. Both seed and labour are thereby saved, and good crops, with much more certainty, are to be depended upon.

Paring and burning the sward of some clayey soils, may be practised with advantage, as the burnt clay will diminish the stiffness of the soil, and render it more

pervious

pervious to water. This may be still more œconomically effected, and, in other respects, with less injury to the soil, by half burning the clay in clamps or in kilns. A preference can only be given in situations where fuel can at a cheap rate be procured for this purpose.

## SOILS,

### *CALCAREOUS OR CHALKY.*

A PURE unmixed chalky soil, like a pure or lean clayey one, is unfertile. The fertility of this soil, like all others, depends on its containing a due admixture of other earths, with the requisite quantity of vegetable or animal matter. A chalky loam, or mixture of chalk with clay, is frequently a very fertile soil, and well adapted to the culture of beans and wheat.

Chalky soils produce a short sweet herbage, and, for the most part, are more proper for sheep pasture than for tillage. There are no soils that receive more benefit from artificial watering, as they are apt at certain seasons to be parched by drought. Chalky soils that produce short sweet herbage, should not in general be broken up, or converted into arable lands, a practice which will be attended with injury to the soil, and loss to the farmer, unless they are cropped with moderation, well manured, and afterwards properly laid down with pasture grasses.

Clay

Clay is the fittest substance to be applied with a view to alter the arrangement of the parts of a chalky soil. Peat is a good application to soils of this nature, which are frequently termed hungry soils, and very deficient in vegetable matter. And as a sufficiency of dung is not to be procured to manure fully every part of a farm, peat may be applied in one or other of the states of preparation already mentioned. Unfortunately for the improvement of chalky soils, neither clay nor peat is to be found but at the extremities or outskirts of the extensive tracts of chalky countries: wherever they are to be had, the application of them should not be neglected. Calcareous soils, which have long been under the plough, contain a large proportion of phosphat and oxalat of lime. These insoluble saline matters may be rendered serviceable to vegetation by alkalis, vitriolic acid, vitriolic neutral salts, (especially if superacidulated) and by pyrituous and aluminous substances. Even green vitriol, which has hitherto been considered as unfriendly to vegetation, will, when in a proper manner applied to soils like this, considerably improve and promote the growth of pasture grasses.

It

It is earnestly wished, that this season may not pass over without a series of experiments, as above recommended, being instituted by the intelligent farmers on chalky soils, and that the result of them be communicated to the Board of Agriculture, for the general information of the country.

The principal disadvantage attending chalky soils, is that of their being too dry and parched at certain seasons; but possibly this defect, when they are under pasture, may be counter-balanced by the more early grass they produce in the spring, as well as the luxuriant herbage that succeeds the autumnal rains.

These observations are offered with deference to the opinion of others, who may have had more opportunities of making remarks on such soils, and of drawing the necessary conclusions.

Chalky soils are peculiarly well adapted to the growth of saint foin, especially when coal ashes can be had as an article of manure. But as in most situations they are not to be procured in sufficient quantities, and as it is probable

that

that such ashes owe their fertilizing powers to the vitriolic salts they contain, it may be a matter of prudence
and œconomy to apply the salts themselves, as a few
hundred weights will suffice for an acre of ground.

By discoveries which have been made in preparing
these salts, they could be afforded at a cheap rate, were
the present high duties on sea salt, and the refuse liquor
of salt works, taken off, so far as might relate to these
purposes. A regulation of such infinite consequence to
the improvement and more complete cultivation of the
lands in Britain, is of itself a sufficient apology for calling the attention of the Legislature a second time to so important an object of relief and encouragement to the agriculture of these kingdoms. Should measures so essential
to the future prosperity of the country be disregarded,
the inhabitants of this Island ought not to be surprized if
France should hereafter take the lead in the cultivation
of the earth as she has lately done in other valuable improvements; to effect which, the total abolition of the
gabelle, or duties on salt, has, or more properly speaking,
will give her agriculture great advantage over the rest of
Europe.

Y                    SANDY

## SANDY SOIL.

THERE is no soil so naturally barren or unfruitful, but that it may be ameliorated by the industry of man. The very extensive improvements made on the barren sands of Norfolk, prove the truth of this assertion ; the present advanced state of these lands has been owing to the consolidating of the surface by a due proportion of clay, or of a marly clay, which generally is to be found at no great distance from the surface. On this practice (as it were) of making a soil, it is necessary to state, that much less expence is incurred, and more benefit received, by adding clay to a sandy soil, than adding sand to a clayey soil. It would require, perhaps, from six to ten times the quantity of sand to diminish the adhesion of the latter, than it would require of clay to consolidate the former. The great difference, in point of expence, must be obvious. Sand added to a clayey soil, in a less proportion than would produce the effect required, would be materially injurious to the staple ; because sand, when thus applied, never fails, at certain seasons, to render the clayey soil more untractable and unmanageable ;

for

for which reason burned clay, or brick dust, has in preference been recommended, as an article to be added to
stiff clayey soils. There are some soils consisting almost
exclusively of sand and sea shells, which are astonishingly
fruitful; small shells may therefore be used where clay
is not to be had, although it is very seldom that sand is
not accompanied by clay at a greater or less depth. In
situations, where the clay lies at too great a depth for
open work, and where props or pit timber are to be
procured at a cheap rate, it may be wrought by shafting
and under-ground mining. This idea is a novel one, but
in some situations it probably might be carried into exe-
cution with advantage. As shells consist of calcareous
matter, lime-rubbish, or *effette*-lime, would, for the same
reason, be of service to sandy soils; although lime opens
a stiff soil, it is found to have an opposite effect on a
loose sandy soil : still there is not any application to such
a soil, so proper or so fitted as marl, with the assistance
of proper dunging, in the rotation of the farm; care
being had not to injure such forced or artificial soils by
a too frequent and an improper use of the plough, the bad
effects of which are beginning to be felt in the County
of Norfolk.

IN-

## OUTFIELD AND INFIELD LANDS.

THE arable land in Scotland formerly consisted of out-field and infield.  The infield, in the treatment it re-ceived, and in its quality, resembles the inclosed culti-vated lands in England ; while the outfield was similar to the uninclosed common field lands in this country. There is reason to believe that this distinction, prior to the date of inclosures, was likewise general throughout England.  It is wearing out fast in Scotland, from the same cause.

That part of the farm, called the outfield land, never receives any manure.  After taking from it two or three crops of grain, it is left in the state it was in at reaping the last crop, without sowing thereon grass-seeds, for the production of any sort of herbage.  During the first two or three years, a sufficiency of grass to maintain a couple of rabbits per acre is scarcely produced.  In the course of some years it acquires a sward, and after having been depastured for some years more, it is again

sub-

submitted to the same barbarous system of husbandry.

The other division of the arable land consisted of the
infield or croft land, to which the whole dung produced
on the farm was exclusively applied.  By this mode of
treatment, these last mentioned lands were made fertile
at the expence of the others; and by a repetition of this
practice for many centuries, a superabundance of vegetable matter has therein been accumulated.  Too great a
proportion of inert vegetable matter causes ground of
this description to be too loose and open for most kinds
of grain; particularly for winter corn, which, by the
alternate changes from frost to thaw, and *vice versa*, is
liable either to be destroyed or spewed out of the
ground.

Black infield mould of this description, especially when
it contains a due proportion of calcareous matter, produces a rich and luxuriant herbage: it should therefore
be kept in a proper rotation of pasture and tillage, and
not in tillage alone, as is still the prevailing practice in
many parts of Scotland.

By

By the dung, and still more so by the *urine* of cattle, lands of this nature, after having been depastured for a certain number of years, will be found to have received considerable benefit, and to have become more fitted for the production of crops of grain. This is principally to be ascribed to the effect which the volatile alkali of the urine has, in dissolving a proportion of the superabundant oxygenated inert vegetable matter contained in the soil.

It has been stated in the preceding part of this Work, that stable-yard dung, by long keeping and exposure to air, loses its saline fertilizing powers, and becomes in all respects similar to peat. The same effect, in part, will take place on stable-yard dung when applied to ground. Under the article of Putrefaction it has been observed, that the soluble and saline part of the dung bears but a small proportion to what is insoluble. By the repeated dungings, during so many centuries, there has been accumulated in the infield lands of Scotland too large a proportion of vegetable matter. This surplus, however, may, by the judicious application of lime, alkalis, and other saline substances, be dissolved, and thus made to pro-

duce

duce abundant crops of grass, hay, and corn; which articles (from the superabundance of vegetable matter the infield lands contain) should be returned in the state of dung to the outfield lands, for the most part as deficient of vegetable matter, as the infield lands contain a superabundance.

From the example here adduced on the situation of the infield lands in Scotland, as well as from facts and chemical reasons, it must appear, that the alternate application of dung, or vegetable composts, and saline matters, to ground, is the most judicious method to preserve the soil in a state of fertility, and to prevent too great an accumulation of unproductive vegetable matter.

It has, in several instances been remarked, that infield land, when under pasture, and not eaten down or sufficiently cropped by cattle, (as sometimes happens on such grounds) or when converted into plantations, ceases to produce the same kind of grasses or herbage it would have produced if it had remained under pasture; and this is observed in young plantations, before the effect produced, can be ascribed to the action of the trees in

over-

over-shadowing the ground. Were an experiment on some rich infield land made, by preventing it from being cropped or eaten down by cattle for a certain number of years, a thickness of some inches of turf or peat would be superinduced ; and the land would not again be fitted for the production of grain, or sweet pasture grasses, without paring and burning, or the application of lime or saline substances, to act upon, or dissolve part of the accumulated vegetable matter. Hence it is that ground should contain only a certain proportion of vegetable matter to constitute it rich and fertile mould.

A considerable extent of ground, of this description, (formerly under a high state of cultivation),has been observed in some of the southern Counties of Scotland, the cultivation of which ground had probably been discontinued from the frequent wars between the two nations.

By fallowing, and the application of lime, such a soil may be made fit for the production of grain. By deep ploughing, part of the under-stratum, containing less vegetable matter, may be brought up and mixed with the

the surface mould; by which means not only the staple will be deepened, but the soil will, by the admixture, be rendered more compact and close; whilst the spungy inert vegetable matter contained therein will be decomposed, and brought into less volume by the action of the lime. On these accounts, fallowing and liming consolidate a soil of this description, although they might at first be considered to produce contrary effects.

Whilst these, or other operations, are going forward, for the purpose of making the infield land give out either in grass, hay, or corn, the surplus proportion of vegetable matter, the farmer should apply the greater part of the dung made on his farm to his outfield land, abolishing, as far as may be, the distinction formerly made.

The former system of management, however much it may now be disapproved of, and made a subject of reproach to our ancestors, was nevertheless the only one suited to their means. At a remote period, there were neither a sufficient number of men, horses, nor extent of capital, in Scotland, to admit the adoption of a different system;

z                                                    and

and probably at a still more distant period, causes equally
efficient operated in England against improving the whole
of the surface.   Our forefathers, therefore, acted more
wisely,  by the application of the whole of the dung, to
.bring  part of the land into a high state of cultivation,
than if they had divided or applied it over the whole
extent of the farms they cultivated, whence little benefit
could have accrued to themselves or to posterity :  whereas, by bringing certain portions only, into a high state of
fertility,  a stock of materials has been accumulated, and
left for their descendants to work upon, capable of repay-
ing to the exhausted outfield lands,  with abundant interest, the vegetable matters originally borrowed.

Poor lands, made highly fertile in former times, by
the  addition  of  dung,  vegetable,  and  animal  matters,  have  received  such  a  mechanical  arrangement
of their parts, as well as chemical qualities, that they
never  can  return  again  to  their  original  state.   They
may,  as  far  as  the  production  of  grain  is  concerned,
be exhausted for a time by injudicious cropping; yet
.they will always produce grass when not overrun by
                                                hurt-

hurtful weeds, and will recover the power of producing abundant crops of grain, by being for some years under pasture. Soils of this description can never be completely injured but by paring and burning, which would dissipate and throw into the air the vegetable, animal, and saline matters therein contained.

PEAT

## PEAT MOSSES, FENS, AND POOR BARREN LANDS IN THEIR VICINITY.

THIS subject follows, with peculiar propriety, the description of infield and outfield lands; and, as the difference which subsists between these lands depends upon the larger proportion of vegetable matter which the infield land contains, in a state very similar to that of peat, it is obvious that peat mosses may be rendered of very great utility in improving the poor barren lands, which are generally found in their vicinity.

Peat, or vegetable matter, should be carried from the peat moss to the poor soil, and the surface mould from the poor soil to the peat moss. By these means two beneficial purposes may at the same time be effected. The quantity of such like, or other earthy matter necessary to be added to a peat soil, to alter the mechanical arrangement of its parts, is to be ascertained by proper trials; and, on the other hand, the quantity of peat requisite to be applied to poor soils will be regulated by the quantity

tity of vegetable matter which such soils may already contain.

The most efficacious method of applying peat to poor barren soils, is to mix it with the urine and dung of cattle; on failure of these articles, with alkaline and other salts; and, lastly, with lime.

Experience can only determine the number of loads or weight of these different preparations, which should be given to an acre of ground. From such experiments as have already been made with the preparation of alkaline and some other salts, there is reason to believe that the quantity necessary to be given at one time would not exceed that of a proper dunging, and that equally beneficial effects would be produced.

If this, on general experience, should be proved, the peat mosses in Great Britain and Ireland will not only afford an inexhaustible supply of manure for the poor lands in their vicinity, but may themselves, by the application of alkaline and other salts, be brought to the highest state of fertility.

Peat

Peat soils, which acquire an unctuous rich clamminess, by the application and action of dung, urine, alkaline salts, &c. in partly dissolving the peat, are the fittest of all soils for the growth of hemp. The culture of this plant would be a source of employment to the inhabitants in the winter, in preparing the hemp for market, when prevented by frost and bad weather from working without doors; and would furnish an internal or home supply of an article so indispensably necessary to a maritime state.

The cultivation of hemp * on peat mosses thus improved, would be found to be an excellent preparation for wheat.

Peat mosses and fens have, hitherto, generally been considered either as nuisances, when in an unimproved state, or as soils of the greatest fertility, when cultivated. These different states regard only the peat moss or fen itself, and have no reference to any consequences that might arise by the application of peat, or of any preparation of it to the neighbouring lands. But as peat has been

---

* Were hemp cultivated on an extensive scale in this country, the expressed oil from its seed might advantageously be applied to the manufacture of soap, of a superior quality to that which is now made of tallow; for which purpose large quantities are annually imported from Russia, and other countries.

been discovered to be capable of being converted into the most valuable of all manures, the importance of the peat mosses in Great Britain and Ireland cannot, therefore, be too much impressed on the minds of the landowners and occupiers of these countries.

From experiments made with alkaline salts and peat, it can be asserted, that the effects of such mixture, weight for weight, are equal, if not superior to those of dung. The usual quantity of dung given to an acre of ground, when it is intended to manure it effectually, is about eighteen tons. This it may, perhaps, receive once in five years. It will require a farm to be well managed, and in high cultivation, to admit of one-fifth of it to be annually so manured.

The best cultivated land, by the return of its own manure, unless when a quantity of vegetable matter has been accumulated, as described under the article of Infield Land, can do no more than keep itself in heart: of course, nothing can be spared for the amelioration or improvement of poor or waste lands. The rendering the inert vegetable matter of p  t mosses and fens serviceable to this purpose, though effected at a greater expence

than

than is at present incurred by any application or dressing
to ground, could not fail to answer the expectation of the
farmer, and must be considered as one of the most valuable improvements that has hitherto occurred in the
annals of husbandry.

The primary step towards improving a peat moss, is
to take off by proper channels the great feeder of water.
This is to be effected by conducting one or more principal drains through the moss, and by water courses on the
solid or dry land, immediately above the level of the
moss, so that it shall not be inundated by the surface
water or springs of the surrounding higher lands, and
shall afterwards only require to be freed from the water
that shall fall on its superficies. This being accomplished,
the intermediate parts of the bog should be drained,
partly by open and partly by covered drains; care being
taken, that they are not made so deep as to lay the moss
or bog too dry; by which the peat, becoming oxygenated,
and thence insoluble, would be incapable of yielding food
to vegetables. By the opposite extreme, unprofitable
grasses and aquatic vegetables are produced. It is therefore an object of great importance, in effecting the drainage,

age, to preserve a full command of the water, that it may be regulated at pleasure.

Should a moss or bog be so circumstanced, as to admit of its being drained through a country which might be improved by a supply of vegetable matter, it would be prudent, previously to making the great or leading drains through the adjacent country or moss, to concert measures for making a proper communication by a canal, so that the remote lands might be supplied with peat from the moss, and the moss supplied with lime, marl, clay, earth, sand, shells, and other materials, from the distant country.

As shell-marl is frequently found under peat mosses, they should be bored in different places, to ascertain if they possess this valuable substance ; or if they contain under the moss a rich clay or limestone gravel: for, in the event of these being found at a moderate depth, the moss or bog may be improved at a much less expence, as the distant carriage of these necessary articles, and the difficulty of carting and laying them on so loose and spungy a surface, would thereby be diminished.

The burning a part of the upper surface may, in some cases, be requisite, to afford, when it is otherwise too difficult to be procured, a due proportion of earthy matter.. Peat ashes will, in such cases, act in making a different mechanical arrangement in the soil: but, near the sides of the bogs where surface mould is to be had, a preference should be given to it.

The alteration in the mechanical arrangement of the soil being effected, the next object is the application of such substances as will bring the peat, or inert vegetable matter, into action. These substances are lime and alkaline salts, which contribute in different ways to the proposed improvement.

Improved peat mosses, bogs, or reclaimed fen lands, are the soils the most productive of luxuriant vegetation, although from this cause they do not in general yield, in this northern and humid climate, heavy and well filled grain. Such soils should be principally dedicated to pasture, and should only be ploughed when, notwithstanding the utmost endeavours, the ground produces coarse or rank grass; but this is in a great measure, or perhaps entirely,

entirely, to be prevented by due attention to the following directions:

To keeping the water in the ditches at a proper level:

To stocking the ground with a due proportion of neat cattle, sheep, and horses; as the one animal will eat the grass which springs up from the dung of the other, and which otherwise would produce tufts of coarse grass:

To folding of cattle on different parts of each field:

To using of heavy rollers:

To top-dressings of alkaline salts, and other saline substances; and also to top-dressings of lime, either by itself, or when mixed with peat or fen mould.

The pasture should always be eaten quite close before winter, excepting such portions of it as are intended for winter food, which likewise should be eaten close off, before the spring vegetation commences: after which

time,

time, it should not be depastured until the grass be
sufficiently advanced to allow of a good bite. As all plants
receive nourishment two ways, by their stems and leaves,
as well as by their roots, the growth of young vegetables
must necessarily be much retarded, when deprived, by
being constantly eaten down, of one of their sources of
subsistence.

As the consolidating the soil of peat mosses is an object
of the first consideration, it is obvious, both on chemical
and mechanical principles, that much cropping, and the
consequent exposure of fresh surfaces to the action of air,
are improper : and that after a few crops are obtained, the
ground should be laid down with meadow grasses and white
clover, and depastured with as much stock as it will carry.
By these means, not only the soil will be consolidated and
compressed, and particular grasses or herbage promoted
according to the cattle so depastured, but the urine and
dung of the cattle, by carefully folding them, will be laid
on in such quantities, as will perform rapidly, and at once,
the effect required; whereas, if the same quantity were
divided over fifty or a hundred times the surface so folded
upon, its operation would scarcely be perceptible, and the
application

application of each year's manure so subdivided, would either be washed away, or its beneficial effects lost, before the further quantity necessary could, in a series of fifty or a hundred years, be added to it. Hence a system of under-dunging, or manuring land, may be said to be nearly equal to no dunging at all: on which account the preference, with great reason, has been given, under all circumstances, to the ancient mode of cultivating the in-field lands of Scotland.

Folding in a proper manner, is particularly recommended for fen lands and peat mosses, as the immediate effect produced by *urine*, is that of dissolving into mucilaginous saponaceous matter the oxygenated peat. Indeed, in all businesses, it is well known, that what is once well done is not to do again. By this judicious mode of proceeding, the chemical qualities and mechanical arrangement of the soil are so altered, that, without the grossest mismanagement, it is impossible it should return again to its former unproductive state. If this be properly attended to, the pasture will never grow coarse, or require breaking up, but will continue to improve the longer it is suffered to remain in that state. There are other rea-

sons

sons for continuing such lands in pasture, viz, the difficulty, from the softness of the soil, of conveying dung
from the farm-yard to the fields; and likewise the tendency which such soils have, by exposure to air, to become oxygenated, and consequently incapable of yielding
the food requisite for the support of vegetables.

In the mode of stocking such pastures, it is further
recommended to keep the fat, the half fed cattle, and
the lean or young stock, in different inclosures, as is the
practice in Ireland, and where grazing is well understood
in England. The fat cattle should only top the grass, the
half fed should succeed those, and lastly, the lean or
store cattle should follow on the same pasture, and eat
the herbage close down; repeating this practice as often
as the fresh growth of grass will permit.

DRAIN-

## DRAINAGE OF THE FENS.

THE loss sustained by individuals and by the public, from the late breaking of the banks, and consequent inundation of the cultivated fens in Cambridgeshire and other Counties, necessarily calls the attention of the proprietors of fen estates, and the Legislature of this country, to a more judicious and complete system of general drainage of that great level. The late destruction of the banks, and loss sustained, may probably be the means of uniting, in one natural and judicious plan for their mutual advantage, the hitherto distinct and opposite interests of individuals and of neighbouring communities. This is most likely to be accomplished by adopting the method of drainage recommended in the Appendix to the very able Report of the Agricultural Survey of the County of Cambridge; by which plan there is reason to believe that the fens might be wholly, instead of being partially, drained: an object of too great importance to be overlooked, or to be thwarted by the confined or mistaken plans of any class of men, or the interested motives of

any

any particular place or borough, with a view to make
such place, or borough, the mouth or outlet, and consequently the shipping port to the fens so drained. To
such objects, the general interest of the many, and the
nation in general, are too frequently sacrificed. It is
not those who propose the best plan of a general drainage, line, and direction of a canal, or rendering a river
or rivers navigable, who carry their point; but it is those
who can procure most friends in the two Houses of Par-
liament. Many judicious plans of canals, and of rendering rivers navigable, have been rejected, and the interest of proprietors of an established *ditch* of a canal,
has been allowed, to prevent the cutting of canals of
much more general utility and importance.

The late accidents which have happened to the fens,
by the breaking of the banks, should lead to an impartial
survey, and consideration of the most judicious method
of draining this valuable country at the lowest possible
level, so as to avoid the raising of the water, as has hitherto been the case, to a much higher level than is said
to be necessary.

On

On this subject it is proper to remark, that exclusively of the main central drain, or drains, carried at the lowest level, and according to the natural and old established course of the waters, sufficient drains or water courses should be made to skirt or surround the whole of the fens cut on the dry or solid land above their level. It is obvious, that by thus guarding the fens from their great and principal feeders which come from the upland country, there would then require to be drained, or to be raised from the fens, only that proportion of water which may fall on their superfices. The Agricultural Surveyor of Cambridgeshire seems to be of opinion, that the general bed of the fens is sufficiently elevated above the level of the sea to drain itself. This fact is of too important a nature not to be fully and minutely *inquired into*, as in the event of its being established, not only the fens lately under cultivation, and now drowned, might in future be more securely and effectually drained, but an addition of 150,000 acres of undrained and unreclaimed fen land, in the County of Cambridge *alone*, would accrue to the agriculture, or the cultivation of this country. Should the fens not be capable of being drained completely by any sea level, and that the water should require to be raised, the most judicious method of proceeding would be still to conduct the water to the lowest level, and which will be found to be nearest to the sea, and then by a sufficient number of windmills, or other engines, to lift it over the great sea bank. Some well constructed fire engines would at *certain times* be of material service to the drainage.

Fire engines employed in the accomplishment of a great and national object, should be exempted from duty on the coal required to work them. The expence of fuel might still farther be diminished, by making use of the refuse small coal made at Newcastle in working the larger and more valuable sorts. This refuse coal may there be had at two shillings and three-pence per London chaldron of thirty-six Winchester bushels, and of which, about 100,000 such chaldrons are annually brought from the pits for the purpose of clearing the underground workings, and are allowed to decay, and perish on bank. Were this quantity of *refuse* coal allowed to be applied, duty free, to such like, or to other important national objects, many beneficial consequences would thence arise; the public revenue could not suffer by exempting.

-empting this at present waste or refuse article from duty for the working of fire engines, or the burning of lime ; still any limited or partial exemption from so injudicious a tax as that on water borne coals, would be far short of the advantages, which, by a total repeal of the duties, would ensue to agriculture, manufactures, mines, machinery, navigation, or extent of shipping, and the health and comforts of the great body of the people in this humid and northern climate. An exclusive duty on coals carried coastways, may be deemed a *prohibition* to the rearing of *seamen*, and a *bounty* on the breeding of *horses*. It is truly astonishing that so glaring an absurdity has not hitherto been corrected. The late requisitions, and very *strong but necessary steps* taken by Government, to procure seamen to defend our Island from foreign invasion, show whether it is to *horses* or to *seamen* we are now to be indebted for our defence. In Scotland, the duty on coals is repealed, and an additional duty on spirits substituted in its stead. In this respect, the inhabitants of South Britain would do well to promote a similar plan of commutation.

The

The draining, inclosing, and properly cultivating the fens and peat mosses in Britain, would, by rearing and feeding a greater number of cattle of all descriptions, allow a greater proportion of the higher and drier lands to be kept in tillage; whence would be produced a greater quantity of grain and animal food. The present inhabitants of Great Britain would be more reasonably and plentifully fed and cloathed, and a considerable surplus would be left either for exportation, or for the maintenance of an augmented number of people.

Population would increase as plenty is secured. The additional produce of the earth would not only feed a greater number of inhabitants, but would provide them with constant employment in the manufacturing of wool, hides, hemp, and flax, the internal productions of our own Island, instead of relying upon a precarious supply of some of these necessary articles from foreign States; and lastly, *emigration, the constant attendant on scarcity, would no longer rob these kingdoms of their only defence.*

WEST.

## WEST INDIA ISLANDS.

BENEFIT TO THE CULTURE OF THE ISLANDS FROM A DUE
ATTENTION TO THE OXYGENATION OF VEGETABLE
MATTER, AND THE SUBSEQUENT SOLUTION OF IT, BY
ALKALIS. AND OTHER SALINE BODIES.

THESE Islands, when first known, were covered
with wood; the lands have since been cleared, and the
soil has been brought under the culture of the sugar-
cane, indigo, and cotton. The crops of sugar, for many
years, after clearing and cultivating the ground, were
very great. At length, by repeatedly cropping and exhausting the soil, the planters are now under the
necessity either of manuring for sugar crops, or of
substituting others of a less exhausting nature for a
certain number of years, until the ground shall recover.
The very abundant crops of sugar at first produced, were
undoubtedly owing to the accumulation of vegetable
matter, in consequence of the Islands having been covered with wool for many centuries previous to their
settlement and cultivation.

The

There can be no doubt that the vegetable matter has in a small part only been expended in producing those abundant crops, and that by far the greater part has become oxygenated and insoluble by the exposure of renewed surfaces to the action of the atmospheric air, in consequence of the *frequent stirrings and hoeings* which the ground has received.

No soil can, in any climate, continue to produce in abundance, sugar, grain, or other exhausting crops, without receiving back in return such a proportion of vegetable matter, in the state of dung or otherwise, as would be equal to the weight of the vegetable matter afforded by the *soil* to each crop. By this, it is not meant, nor would it be possible, that there should annually be returned to the ground as great a weight of vegetable matter as shall be equal to the weight of the preceding *crop*, it being only necessary to return as much as would be equal to the proportion of the vegetable matter furnished by the *soil*.

By much the greater part of the vegetable matter now existing or remaining on the surface of the earth is indebted to the aery form fluids or gasses, and to the de-
compo-

composition of water in the process of vegetation, for the matter of which it principally consists. To these two causes the accumulation of vegetable matter on the surface of the earth is principally to be ascribed. It would not be an easy task to ascertain the different proportions of the food of plants supplied from the soil, or from the air, and the decomposition of water. It must be admitted, that the constituent principles of the vegetable substances con-tained in ground long under cultivation, have, in a great measure, been supplied to such vegetables, when growing, by air and water. This must be evident from the large proportion of vegetable and animal matter (under all states or systems of cultivation or cropping) annually taken from the soil, consisting of grain, or animal sub-stances, such as flesh, milk, butter, and cheese; the dung and urine of cattle dropped on the roads, (where it is totally lost) and the large proportion of the food of animals, thrown off by insensible perspiration and breathing: to all which must be added the quantity of extractive soluble matter annually washed away by the rains, and carried into the sea. Had not the benevolent Creator of all things established processes for supplying such an unavoidable waste of vegetable or other matters,

the

the surface or soil of all countries would, in process of time, have become barren and unfertile. On the whole, there can be little risque in asserting, that the quantity of vegetable matter in soils under a proper system of cultivation and dunging, is, notwithstanding, the above abstractions, probably on the increase. The great quantity of saccharine and other matter yearly taken away, without being returned to the soil, and the unfitness of the hard or woody part of the sugar-cane for the food of cattle, evidently prove that the sugar plantations must suffer annually a considerable degree of deterioration ; and that from so great an abstraction, as well as from the process of oxygenation, the soil must at length cease to produce crops of sugar. Such effects have in part been already experienced in Barbadoes, and in the Islands first settled and cultivated.

It is apprehended, that the less productive power of the soil of some of these islands, is more to be ascribed to the state in which the vegetable matter is now in, than to its having been exhausted by cropping. Should this conjecture prove well founded, and that the soil still contain a large proportion of oxygenated vegetable matter,

the

the surface may again, for a time, be rendered fertile by the application of alkaline salts and other saline matters: repeated applications would undoubtedly exhaust any soil, and as such these should only be applied alternately with dung or vegetable matters. Every judicious planter keeps a sufficient number of live stock for the purpose of making manure, and sets apart a due proportion of his plantation for producing proper food for their maintenance. It is necessary that cattle should be kept in great numbers, to provide in some measure for the waste annually incurred in the sugar grounds, still there will be a deficiency of manure from the stock so kept, to preserve the sugar-cane lands in high condition.

An idea is here suggested of supplying that deficiency, founded on the principles and facts already mentioned, towards which the following, as it respects the cultivation of the sugar-cane, is offered as a more full explanation.

An acre of ground under canes is generally reckoned to produce an hogshead or about twelve cwt. of sugar, besides the molasses, &c. The tonnage of ships necessary to convey such bulky articles to Britain, must be

c c                                          very

very considerable. Those exported, consisting chiefly of valuable manufactured goods, are of small weight when compared with the weight of the sugar, &c. imported; and of so little value is the freight of outward-bound ships, that they are frequently either entirely laden with bricks, lime, &c. or partly so by way of ballast: in which last situation most of our West Indiamen perform their outward-bound voyage.

As dung is also said to be sent from Britain to the West Indies, this circumstance, (together with the chemical remarks and observations on the nature of peat) has pointed out peat as an article capable of being sent in certain states of preparation to the West India Islands, with a view to supply the annual consumption and deficiency of vegetable matter, which may take place in the soil by the cultivation of the sugar-cane.

This idea originated from knowing that the soreline acid, the acid most abundantly contained in peat, and the acid of sugar, were identically the same.

There

There can be no doubt, that when peat is rendered completely soluble, and thus fitted to promote the growth of plants, it will, when applied to the culture of the sugar-cane, afford those substances which constitute sugar; when these, by the process of vegetation, are afterwards combined and united in due proportions.

To send cargoes of peat in an unprepared state to the West Indies, would be the height of folly and absurdity, as no vessel could carry enough of so light a substance even to ballast her. But as peat, when dissolved by alkaline salts, and afterwards dried, may be brought to the consistency of a solid dry gum, equal in weight perhaps to forty or fifty lb. per cubic foot, the objection to its lightness would thus be remedied, and it might be exported to the West Indies at a low return freight, or as ballast.

If there be any accumulated masses of vegetable matter, or peat, in the West India Islands, this manure might certainly be prepared there, at a cheaper rate than in Britain or Ireland.

The

The improvement here suggested by the application of peat as a manure to sugar plantations, is perfectly new, and as such, with a certain set of men, (whose opinions are not worthy of attention) the proposal may be subjected to the epithet of a *scheme*, an appellation now generally bestowed on all designs, whether the discovery be valuable or useless. Mankind seem at this moment to detest all alterations, excepting in such matters as are the most difficult and dangerous, viz. the *laws*, the form of *government*, *religion*, and the *scheme of society* of a country, in which the most subordinate ranks feel themselves as competent to alter or amend, as a GROTIUS or a MONTESQUIEU.

## CULTIVATION OF SOREL, WITH A VIEW TO THE PRODUCTION OF OTHER MORE VALUABL E CROPS.

IT does not appear that any farmer has cultivated, or that any writer has recommended the growth of those plants to be promoted, which seem indigenous to any particular soil, with intention of rendering such plants of use in the future produ&ction of grain, or the rich herbage upon which cattle feed.

It is no uncommon pra&ctice to sow buck wheat, tares, and other green crops, for the sole purpose of ploughing them in; thus providing the ground with a proportion of *fresh* vegetable matter, at times when other manure cannot be procured, and also promoting the dissolution of the inert vegetable matter contained in the soil by the stagnation of air, and by the retention of humidity, occasioned by the close cover and shade these crops af-ford.

Judicious as this process in many cases may be, of en-couraging the growth of certain vegetables, with a view

of

of promoting, by their subsequent destruction, the future growth of others, yet the beneficial effects of this system of cultivation, by due attention to the vegetables which certain soils have a tendency to produce, is to be extended much further than most farmers are aware of. To attempt making such soils produce, without chemical acids, other vegetables more serviceable to men and cattle, would be premature, as it would be an endeavour to force nature to productions for which she is not as yet prepared.

Soils not calcareous, containing much inert vegetable matter or peat, have a tendency to produce wild sorel, a plant considered in general as an indication of the want of fertility in the soil. This is certainly correct, if the fertility of the soil is only to be estimated by the use or value at market of the crop, but not as it respects vegetation itself; for a soil of the above description often produces a most plentiful crop of sorel. In this case, as it applies to the further improvement of the land, the growth of sorel should as much as possible be encouraged, even by sowing the seed for this especial purpose. The vegetation of this plant is no doubt promoted in the

the soil by the oxalic or soreline acid, formed by the combination of oxygen, or pure air, with the basis of the soreline acid contained in the vegetable matter of the soil ;
and so long as the vegetable matter remains in a state
fit to become oxygenated, it will have a tendency to
promote the growth of sorel. It has been stated that
the juice or salt of sorel is a superacidulated neutral salt, consisting of the vegetable alkali and the
oxalic acid. This superabundant acid is inimical to the
growth of grain, or of such vegetables or grasses as constitute the food of most animals : but which tendency
in the soil, and injurious consequences, are to be corrected by the application of different substances, viz. by lime,
by chalk, by magnesia, by alkaline salts, and by paring
and burning.

Lime will combine with the acid of the sorel, and form
an oxalat of lime, which is insoluble : as such it should
only be applied in such small quantities as will neutralize
the acid in the soil, or the superabundant proportion of acid
contained in the sorel; so that the other component part
of sorel, viz. the oxalat of potash, may not be decomposed by the superior affinity which the oxalic acid has to
lime ;

lime; in which case, the alkali would be disengaged. No injury will arise from the application of a superabundance of lime, provided that the soil contain a still greater proportion of vegetable matter; in which case, the alkali disengaged by the lime would act upon the vegetable matter, and form a saline substance, similar to that which the superabundant use of lime had decomposed.

Ground of this description, to which lime has been applied, will no longer have a tendency to promote the growth of sorel in preference to other plants; its next spontaneous growth will, probably, be chickweed, which is a certain indication of its being in a state fit to produce grain or other crops.

Magnesia has a greater affinity with the oxalic acid than alkalis have, so that by the addition of earths, containing magnesia, to ground producing a crop of sorel, the acid will not only be neutralized, but the oxalat of potash, the other component part of sorel, will likewise be decomposed. By this means the alkali will be disengaged, and put into a situation to act upon, and dissolve the inert vegetable matter contained in the soil. The salt

formed

formed by the combination of the magnesian earth with the oxalic acid, will, as well as the vegetable matter dissolved by the alkali, be found to promote vegetation in a very great degree; hence magnesia, by forming with the oxalic acid a soluble salt, has an advantage over lime, which forms with the same acid a salt that is nearly insoluble, but capable of being brought into action by methods previously stated.

By the application of alkaline salts to sorel, there results a salt fully neutralized, which highly promotes the vegetation or growth of more valuable plants and grain.

When neither magnesia nor alkaline salts are to be procured, and where it is not thought proper to make use of lime, the thinly paring and burning the sward, consisting of the plants of sorel and their roots, may be performed with advantage, as a *large* proportion of alkaline salts will be procured for the purpose of dissolving a correspondent proportion of the inert vegetable matter contained in the soil. This operation of paring and burning is only recommended as a last resource,

and

and in performing it, great care should be taken that as little of the surface mould as possible be consumed by fire.

The application of the different substances here recommended, or the operation of paring and burning, should take place only at the time when the crop of sorel is in the greatest luxuriance.

CON-

## CONCLUSION.

THE multiplicity of subjects connected with the main object of this work, open so wide a field of enquiry and discussion, that it is with difficulty the Author has restrained himself from exceeding the bounds proposed; " *Cuncta me non dicturum sed quædam.*" More might have been said on the practical part of husbandry; but, unluckily for science, too much has already been written on that subject, and absurd theories have been too often blended with practice. Beside, it has been taken for granted, that the persons for whose use and benefit this chemico-agricultural Treatise has been composed were well acquainted with the general operations of husbandry.

As the main object of this work is to promote the more complete and extended cultivation of the soil, nothing which may retard or advance this important object, can be deemed foreign to this publication; as such, the form of government, laws, particular taxes, general system of taxation, manufactures, habits, and manners of different classes of society, orders, or professions of men, ought not to be considered too great a digression.

Many of these produce consequences inimical to the interests of agriculture. The Author's natural turn has led him for many years to direct his thoughts to such objects; but as on several of them he could not offer an opinion without severe animadversions, he has chosen rather to be silent, than to probe sores which require, *at this time*, to be touched with the most gentle hand; at a time when the *morbus Gallicus* has succeeded to the *Negroe rabies*, in the same manner as this distemper succeeded to the *hydrophobia*, or *canine madness*, which for some years was so prevalent throughout Great Britain.

There are few who have viewed in a stronger point of light the necessity of a reform; but it is principally a reform in the conduct, hearts, and pursuits of individuals, as stated in the parochial resolutions of Culross, in November 1792.

The

The reformers, and their views of reforming, fall under some one or other of the following heads.

The manœuvres or plans of one political party to supplant another.

The intentions of the worthless and desperate to overthrow, *in toto*, the form of government, religion, laws, and present scheme of society of this country, and to substitute in its stead the *French* scheme of society, which has already been exhibited on the *theatre* of Paris, and in other parts of that unfortunate and distracted kingdom.

There is another set of reformers, who are worthy and upright men, who view with regret the glorious and unequalled government of Britain slowly *overturning*, by the abuses and corruptions, which, proceeding from avarice and politically mercantile venality, now deluge this nation. These good and worthy people, like the daughters of ISRAEL, " with their harps hung on the willows of " Babylon, weeping over their beloved Jerusalem," have bewailed, in too strong and unguarded terms, the state

of

of the times ; and although good men in their own pri-
vate characters, perhaps may be ill calculated to act the
part of Reformers of States, or of the pursuits or vices
of individuals, unless by their own upright conduct and
example.

Unfortunately this last class of reformers have been
led for a time to make common cause with the two
other classes, whose views have been proved to be very
different; but now there is reason to believe, that all
those of religion, character, or worth, who were reformers *upon principle*, have for the *present* dropt all intention of reform ; and, instead of exciting the desperate
and needy to acts of sedition and violence, do now, like
LOT, pray unto the LORD to spare the country on account of the righteous persons it still contains, and of
which there are many to be found, but more particularly in the middling walks of life.

It is the present too prevailing philosophic infidelity
and departure from the religion of CHRIST, foretold to
happen in the latter times, which has given rise to those
turbulent passions, which agitate the minds of men,

and

and which now convulse and shake the Continent of Europe to its centre. The French, and other freethinkers, particularly VOLTAIRE, the supreme God, or JUPITER, of their heathenish pantheon, have done irreparable injury; and sorry is the Author to observe, that too many gentlemen of the learned professions, as well as those who dedicate their time to literary and philosophical pursuits, are actuated by a similar infidelity, and a desire to explode a religion which raised the minds and virtues of the primitive Christians above those of other men.

Leaving, however, a subject on which it is unnecessary to say more, than to regret that it is so, the Author will now proceed to touch on some points which have appeared to him more proper to be stated in his Conclusion, than in the body of the work.

The first that presents itself, is the benefit that would arise to farmers, breeders, and others, were they allowed to malt the grain now given to horses and other cattle. The following extract from a very valuable work, viz.

viz. Mr. EDWARDS's History of the West India Islands, and which only was perused since the other parts of this work went to press, corroborates, in the strongest manner, the benefits that horses and cattle would receive, were the grain they are fed with converted into a sweet or saccharine substance.

" The time of crop in the Sugar Islands, is the season
" of gladness and festivity to man and beast.  So palata-
" ble, salutary, and nourishing is the juice of the cane,
" that every individual of the animal creation drinking
" freely of it, derives health and vigour from its use.—
" The meagre and sickly among the negroes exhibit a
" surprising alteration in a few weeks after the mill is set
" in action.  The labouring horses, oxen, and mules,
" though almost constantly at work during this season,
" yet being indulged with plenty of the green tops of
" this noble plant, and some of the scummings from the
" boiling-house, improve more than at any other period
" of the year.  Even the pigs and poultry fatten on the
" refuse.  In short, on a well regulated plantation, under
" a humane and benevolent director, there is such an ap-
" pearance, during crop time, of health, plenty, and busy
chear-

" cheerfulness as to soften in a great measure the hard-
" ships of slavery, and induce a spectator to hope, when
" the miseries of life *are said* to be unsupportable, that
" they are sometimes exaggerated through the medium
" of fancy."

On this there is the following note.—" He (says honest
SLARE the physician) that undertakes to argue against
" *sweets* in general, takes upon him a very difficult task,
" for nature seems to have recommended this taste to all
" sorts of creatures; the birds of the air, the beasts of
" the field, many reptiles and flies, seem to be pleased
" and delighted with the specific relish of all sweets, and
" to distaste its contrary. Now the sugar-cane, or sugar,
" I hold for the top and highest standard of vegetable
" sweets." " Sugar is obtainable in some degree from most
" vegetables, and Dr. CULLEN is of opinion, that sugar
" is *directly* nutritious. There is also good reason to sup-
" pose, that the general use of sugar in Europe has had
" the effect of extinguishing the scurvy, the plague, and
" many other diseases formerly epidemical."

These

These authorities experimentally strengthen and cor-
roborate the **Author's** opinion, that horses and cattle
would receive more benefit, by the grain they are fed with
being previously converted into saccharine matter (as is
the case with malted grain) than by being fed with raw
grain, containing no such sweet or saccharine matter: the
expence of malting, exclusively of the duty, is but a trifle,
not exceeding sixteen-pence per quarter.

The duties on malt, used for brewing and distilling
should not be allowed to deprive working horses and
cattle of their share of the *sweets* and comforts of life,
not only enjoyed by the negroes in the West Indies, but
by the pigs, poultry, horses, oxen, and mules, whose
festive board, and *solatium,* (from the *said to be* slavery
and miseries of life) happens only annually, while the
Author of more extensive benevolence wishes that the
*sweets* of life shall in this country be administered daily to
all these animals.

Convincing as experiments may be, and the arguments
in favour of procuring to cattle in this country a *substitute* for the *sweets* of the West Indies, still, from the too
rigid

rigid attention of Ministers to the system of finance, little hopes are to be entertained, that a permission to malt grain for cattle will be obtained, unless it can be made an object of immediate taxation.

Such considerations are permitted too frequently to counter-balance or weigh down other objects of *much greater national importance*; which, if properly fostered and cherished in their infancy, might become resources of national revenue.

These remarks will be confined to the malt-tax, the salt-tax, and the coal-tax, although there are *many others* which tend to repress and cramp the exertions of individuals, and to limit the extension of several valuable branches of manufacture, and of which, were it not for these restraints, an unrivalled monopoly would be secured to this country in most of the foreign markets. By the plans now proposed, the revenue will not suffer; on the contrary, it will receive a considerable increase. The Author hopes that Government and individuals will candidly and deliberately weigh what he has to propose, in his new assumed capacity of a Revenue Officer, or Suggester of

F e 2                    Taxes,

Taxes: an employment which, as it is the *first* he has ever filled, so it is the last he would have thought of filling under Government.

The plans of taxation, and facts upon which those plans are founded, come now to be stated.

There are reckoned to be in Britain ˙ ˙          millions of acres under annual culture by the plough; and supposing each horse to cultivate        acres annually, the number of horses requisite for tillage will amount to.

To keep up this stock, it would require an annual supply of at least a sixteenth part; and as horses are not put to work in general until three years old, three sixteenths fall to be added to the number of horses kept for tillage; at which rate the number would amount to        : to which are to added the horses belonging to the cavalry, those employed in carts, wains, waggons, collieries, mines, machinery, manufactories, and other works; and also horses for posting, carriages, the saddle, hunting, and the turf, &c. &c. These are reckoned to amount at least to one third of what is kept for tillage, so that the whole number of horses kept in Britain.

Britain cannot be less than                    and, at a duty of
five shillings on each horse, mare, or gelding, the whole
would amount to £

As to the neat cattle in Britain, their numbers may be
ascertained, with tolerable precision, according to the following calculation.

There are said to be slaughtered annually in London
and its neighbourhood,              thousand heads of cattle; and as its population is reckoned to be one eighth of
the whole kingdom, but say one ninth, the number of
beeves slaughtered annually in Britain should amount to
              from which one third is to be deducted, from
the greater proportional consumption and waste of animal food in the metropolis. The total number of beeves
slaughtered would therefore be              which mul-
tiplied by four (sooner than which age, cattle are never
brought to market) the total number of cattle of all descriptions, at one time in the kingdom, will amount to
              which at a duty of 1s. per head, for per-
mission to supply them, when necessary, with malted
                                                    grain,

* The blanks have been left to be filled up by the reader, with a view to secure
  .t . tcnaion on his part to the subject, and from a desire to shun
     .ore of figures or calculations..

grain, would produce an annual revenue of £
and the total sum of the duties on horses and cattle* would
amount to £          in addition to the present tax on
malt for domestic consumption, breweries, and distilleries.

Individuals will probably object to this tax; but as the
Author is fully persuaded that such objections, on their
part, can only proceed from their being unacquainted
with the advantages to be derived from a permission to
malt grain for the use of horses and cattle, some pains
shall be taken to obviate such objections; and as it may
be a considerable time before the proposed duty take
place, *if ever* it should, individuals of the Board of Agriculture, or others, may experimentally satisfy themselves,
at the *present duties* which should be reckoned an object of
*no moment* when put in competition with an object of *such
importance* to themselves and to the country in general.
On which head the following statements are offered.

Farm and other horses, when kept even with the most
rigid œconomy, are fed with oats, or other grain, during
seven months, or thirty weeks, of the year; there are
many

* A tax on horses and neat cattle is paid by the inhabitants of Virginia and
Kentucky.

many farm horses that are fed with grain all the year, as is the case likewise with carriage, waggon, post horses, and horses for the saddle. A bushel and a half to each horse per week, is not too great an average allowance for these thirty weeks, this amounts annually to five quarters and five bushels, and at sixteen shillings per quarter, to the sum of 4l. 9s. 7d. Of this expence, at least one fourth will be saved by feeding the cattle on malted grain, but say 20s; from this will fall to be deducted a duty of 5s. to Government, and the sum of 8s. as the charges of malting; still there would be an annual saving on each horse of 7s. But this is a trifling object in consideration of the *higher condition* and order horses would be kept in, fresher and fitter for work, and *freer from many diseases* to which they are now liable.

The benefits that neat cattle would receive at the small tax of 1s. per head are as follow:

By a supply, when necessary, of a small quantity of malt to year-old calves and young stock, they would, when green food is not to be procured, be kept open in the belly, and the *costiveness, binding of the hide*, and *biliary obstructions* which at that season they are liable

to,

to, would by these means be prevented. It should be given properly mashed with hot water, when by the state of their fæces and hides they appear to require it. Too little attention is paid by rearers and breeders of cattle to these necessary precautions.

The next application for which malt has been re-commended, is the clearing away *biliary obstructions*, and opening the bodies of oxen and other cattle, previously to supplying them with the food with which they are in-tended to be fattened; especially when such food is not sufficiently of an aperient nature. And lastly, malted grain, when dry, but more particularly so when mashed, has been recommended to be given to milch cows, to make them yield a *greater quantity* of milk, *richer* than that produced, and *free from the bad flavour* which milk is apt to acquire, by feeding cows in the winter time on cab-bages or turnips.

On the whole, a tax of one shilling per head on neat cattle, seems but a *trifling object* to those who may pay it, when the *benefits from the indulgence* granted shall be considered. And here the remarks on this subject will conclude, by recommending the making of such proper

proper trials of the comparative benefit of the use of malt and unmalted grain, as will fully establish the advantages attending the former practice.

It may be said, that according to this plan, the present malt-tax will be evaded, and that persons would supply themselves for brewing with the malt prepared for feeding their horses and cattle; but this is to be prevented by allowing oats, peas, and beans * *only* to be so malted, in a barn entered for that purpose, under the survey of the Excise Officer of the district, attaching a high penalty, with forfeiture of the grain, for malting any other kinds than such as by law are so allowed. Neither from peas nor beans can a proper *palatable* fermented liquor be made, and the use of oats would be extremely unœconomical, from the small weight of kernel in proportion to the weight and price of the grain.

Lastly, Should these arguments and reasons not be thought sufficient to induce the adoption of the arrange-

F f ments

* Perhaps the malting of peas and beans may not be found so requisite, or so advantageous as the malting of oats, and that all the purposes for feeding of cattle may be obtained by breaking and crushing them in a mill, especially when given mixed with malted oats.

ments here proposed, a total abolition of the present malt-tax seems the only alternative, adopting in part the substitutes already mentioned, and in lieu of the present tax upon malt, an additional tax upon the brewer, with an equivalent tax on all families who brew for their own use, and an additional tax on spirits.

Having stated thus much on a subject, which appears of such importance, and which is likely to be earlier attended to in France * than in this country, a few words may be proper on a plan for the purpose of entirely doing away the present salt duties, and substituting such others as shall fall on individuals in proportion to the benefit they may receive by the use of salt; particularly on those, who, from the high price of it, are now debarred its use in agriculture, the feeding of cattle, and in manufactures.

It is believed that Government would willingly allow the exemption of duty on salt used for these purposes, provided the revenue were not to suffer by the application of such duty-free salt for domestic consumption. Here is the difficulty which has stopped the progress of Ministers

* A tax on malt made no part of the . . . . . . . . . . . . . . . . . . . . . . . . . coun .

nisters and others on this important alteration, whilst to the Author it appears capable of an easy adjustment.

The first object to consider, is the duty per ton which could be afforded, or which would be reasonable to exact for salt applied to the three purposes above mentioned.

This duty should not exceed one pound per ton, or sixpence per bushel, which should be all the tax exacted on the commodity itself. The Author gives it as his opinion, that the *extended* consumption of salt for the purposes already mentioned, as soon as the benefits derived from the use of it were generally known, would, at a duty of one pound per ton, be fully equal to the duties *now collected*; but as it may not be prudent to trust to speculations, *de futuris contingentibus,* it is suggested, that the deficiency of duty which would arise on its present consumption, be made up by a tax on those who should receive a benefit by the unrestrained use of it, such as farmers, feeders, and owners of cattle, and certain other classes of consumers of salt. Householders should only pay for domestic consumption the 6d. per bushel, by which means they would be supplied with salt 4s. 6d. a bushel cheaper than they are at present. At this rate, the

duties

duties on salt for domestic purposes would not exceed 60,000l. a year; but the difference is to be made up by a similar tax to that recommended under the article of Horses and Cattle, together with a duty of 3d. in the pound on the rent paid by each farmer or occupier of ground, and 10s. a ton on the salt rock exported to foreign countries: according to which plans, the salt duties would amount to as follow:

To      tons of salt for domestic consumption, at 1l. per ton   -   -   -   -    £

To      tons of rock salt exported to Ireland and abroad, at 10s. per ton     -     -

To duty on      horses,    -    -

To duty on      neat cattle,    -

To 3d. in the pound on     rent of land

—————

—————

which sum would exceed by       the whole net amount of the salt revenue of Britain.

But the increase of the revenue would not stop here, for as soon as the valuable application of salt for agriculture, and the feeding of cattle, was generally known

and

and practised, the annual consumption of salt, for these purposes only, would at least amount to        tons more. So that the whole revenue from salt would amount to £        Should this be a greater sum than Government would choose to exact, they may limit, or do away all the duties now mentioned, excepting the duty of 20s. per ton, as soon as the consumption of Britain for all the different purposes shall, at the duty of 20s. a ton, amount to the net revenue at present collected.

The tax upon water-borne coals now claims attention. Some remarks have already been made on that subject in page 194, but not so fully as the importance of the subject merits.

The coasting coal trade of Britain is said to employ        vessels, on an average, each of        tons, navigated by *        seamen, most of whom are apprentices; as such the coal trade is reckoned the principal nursery for British seamen. There is no coasting or carrying trade in the world, of which the cargoes, or articles carried,

* For the reasons already given, this blank, with those which precede it, are left to be filled up by the reader.

carried, are so disproportionate, or of so small a value, when compared with the freight and capital employed in shipping; consequently there is no trade which employs so many seamen in proportion to the value of the cargoes. Profitable or judicious branches of trade should by every means be encouraged, especially those in which the persons employed are the protectors and defenders of their country.

No person, who has the prosperity and defence of this Country at heart, but must wish, that that trade should be increased which promotes ends so truly to be desired. It must be obvious that such objects are not to be attained by laying duties on an article when water-borne, of which the carriage by sea should as much as possible be promoted.

To what further extent the coasting coal trade might be carried, is not an easy matter at present to predict; it is capable not only of being increased to a very great degree, by taking off the present duties, but may be increased in a much greater degree, by the extended culture

culture, and consequent population of the kingdom: objects, which, *from necessity*, are now in a very strong manner awakening the attention of individuals. The Author does not think he hazards an opinion, when he says, that the coasting and export coal trade of Britain may be doubled within ten years from the present date; but this is not to be effected without a total repeal of the duties on coal carried coastways.

It is not an easy matter to find a substitute for this tax, at least such a one as would meet with general approbation. Dwelling-houses are already taxed in several ways; first, in all the materials of which they are built; and, secondly, by the land-tax, window-tax, and tea commutation tax: still were there any other tax which could with propriety be levied on that species of property, it would be the coal tax, from the necessity all persons are under in this northern climate of making use of fuel.

One of the many advantages attending such a commutation would be, that it would embrace and *include* the inhabitants of the counties where coal is wrought, who

who at present, in that respect, contribute *nothing* to the exigencies of the state, and who in general, from the saving of carriage, freight, and duties, are supplied with coal at a *third* or a *fourth* of the price paid in other parts by the subjects of the *same* kingdom, than which nothing can be more *partial, unwise,* and *unjust.*

To a general tax, or regulation of this kind, the inhabitants of the coal countries could not with reason object ; nor could any objections be made on the part of the great consumers of coal in different branches of manufacture; because these, as well as fire engines, would be exempted. To make the proposed tax bear equally in proportion to the wealth of individuals, it should be in proportion to the number of windows or chimneys contained in each house. All who consume coals should bear their proportion : none should be exempted but those who receive assistance from the parish, or of whom the Churchwardens certify that they are persons in indigent circumstances.

As the Author is not furnished at present with the amount of the window-tax, nor the commutation

tion tax on tea, he cannot on this head, enter into further detail; and will only add, that were the tax to fall heavier on that species of property than he is aware of, it might be proportionably diminished, and the deficiency of the tax be made up by an additional one on ardent spirits, or on some other proper object of taxation.

Having concluded the suggestions towards the repeal of the three taxes which are the most repressive of individual and national exertion, reference will now be had to Mr. EDWARDS's description of the soils best adapted for the culture of the sugar-cane, in his History of the West Indies, wherein he confirms the opinion of the Author, that soils may receive considerable injury by too frequent ploughings, stirrings, or exposure of renewed surfaces to the action of air and sun, especially where the sun's rays are so very powerful as they are in the West Indies. By these Extracts it will appear, that (with few exceptions) the soil in the West Indies, best adapted for the culture of the sugar-cane, is a rich, deep, *black* mould. There is no soil of that colour that does not contain carbonaceous or *vegetable* matter. Such

soils

soils are peculiarly fitted for the application of *alkaline salts,* that the vegetable matter which had become oxygenated and inert, by frequent culture and exposure of fresh surfaces to the air, may by these means be dissolved. Notwithstanding the fertility of this black soil, it still requires the aid of manure or *urine,* as appears by Mr. EDWARDS's observations on the soil of Barbadoes; and in his remarks on the soil of Jamaica, he says, " *the urine of cattle is the best of all manures* ;" which practically corroborates the Author's theory on the action which the *volatile alkali,* contained in urine, has in dissolving *oxygenated peat, or oxygenated vegetable matter,* and of which, so large a proportion is contained in many of the soils in the West Indies. The prevalence of this *black,* or *vegetable soil,* supercedes the necessity of sending to the West Indies the preparations of peat formerly recommended, this soil requiring only *alkaline salts* as a substitute for the *urine* of cattle, which is by no means to be had in such quantities as the ground *would require,* to insure abundant crops.

Mr. EDWARDS's remarks are very judicious as to substituting coal for boiling the sugar, and the distillation

of

of rum instead of the *cane trash* now used as *fuel:* this cane trash, by being kept a due time in heaps, and afterwards mixed with alkaline salts, might be *returned to the land as manure,* instead of being *dissipated or thrown into the air* by combustion.

The different extracts from Mr. EDWARDS's work here alluded to, are subjoined in the Appendix.

The Author hopes, that by candid readers allowance will be made for the imperfections necessarily attendant on all works of this or any other kind, when sent to the press as composed, without due time having been taken either to revise the composition, or to correct typographical errors, several of which occur in this work, and a list of which is subjoined to the Table of Contents.

The public are requested to view the preceding Treatise more in the light of suggestions and hints, than a full and complete treatise on the different subjects. The Author requests permission to correct an expression which had very inadvertently crept into the Preface, page 8, wherein he says, " The Author flatters himself

" that

" that his labours will be found to open a field of expe-
" riment, of chemical reasoning, and of the practically
" useful, applicable to agriculture, of which that science
" had hitherto been thought incapable." His meaning
was, and it should have been so expressed, " of which
" that science has to too many appeared incapable : " for
persons of reflection and understanding must admit, that
chemistry, defined in the Introduction to this Work, " to
" be a knowledge of the properties of bodies, and of the
" effects resulting from their different combinations," 
cannot but be necessary to the proper understanding and
bringing to perfection, any art, science, or occupation,
wherein *matter* is to be operated upon."

Attacks upon the theories and opinions of other writers
have been carefully avoided, the Author's views not
being to court argument or dispute, but to convey such
information as appeared to him might be useful; nor
would he have departed from this line of conduct in the
following remarks respecting the action of lime, had he
not considered it as being conducive to the interest of
individuals, and to agriculture in general, to combat an
erroneous theory, to which many have subscribed, on
the

the authority of others who have written on this subject.

By these writers it is asserted, that *dung* contains *oil:*
and to this oil the application of lime is recommended,
with a view to render it soluble in water.

No expressions in chemistry or in agriculture have been
so injudiciously made use of as those of sulphur and of
oil. By the word sulphur, brimstone is to be understood;
and b the word oil, those smooth unctuous substances
capable of being *inflamed* or *burned,* produced in the bodies
of animals by the process of animalization, and in the
seeds and kernels of fruits and plants by the process of
vegetation. To which are to be added bituminous oils,
and empyreumatic oils, obtained by the distillation of
animal, vegetable, and some mineral substances, such as
fossile coal, &c. to none of which the juice of dung or
dung-oills bears the smallest resemblance; on the contrary, it will be found to be a mucilaginous neutralized
saline extractive liquor, whence no oil, either from it or
from dung, can be procured but by distillation, or the application of fire; in which case oil cannot be said to be
disen-

disengaged, but *is really formed* in the process, in the same manner as oil is procured by distillation from *mucilage* or *gum*.

Attempts have been made to classify manures, earths, and other substances, under the term of active and passive. Lime is termed active, *i. e.* it is said to possess a power of acting upon other substances, and of making these substances produce or give out to vegetables their proper food. By such theories the effects of lime are carried still further, by ascribing to them the power of rendering oil soluble in water. These oils are termed passive, and are *supposed* to be contained in dung. That such oils *do not exist* in dung, must be maintained, until a *true inflammable oil* is procured from dung by expression, or by some process different from that of fire. What *sort* of oil is meant, the Author is really at a loss to discover; it cannot well be supposed to resemble sallad oil, nor any oil to be had from perfumers, apothecaries, nor the oil shops!

It is known to every well informed chemist, as well as to soap-boilers, that hot lime does not produce, with animal fat or oil, or with the expressed oil from vegetable

seeds

seeds or kernels of fruits, a permanent soap, or saponaceous emulsion; for although such a sapo seems to be
formed, still it keeps only in that state until the lime sa-
turates itself with fixable air, after which a separation
takes place, and the oil comes again to the surface of the
liquor. This proves that lime is not endowed with those
powers ascribed to it, in rendering oil soluble in water.
Admitting, however, (which is not the fact) that the sapo
made by lime and oil was a true and a permanent one,
still the oil, strictly speaking, could not be considered as
in solution, being only mechanically separated and divided in the water, and not chemically united therewith,
as is apparent by its colour; for liquors wherein chemical
union exists are colourless.

The only effect that lime has upon *gross* oil is to deprive it
of that principle which constituted it such, and to reduce
it to the state of an oil more fluid, and more resembling
animal or other oils attenuated by frequent distillations.

Lime certainly does no more than deprive the gross oil of
the acid, on which depend its thickness and consistency,
the lime combining with the acid in the same manner as in
the

the process of soap-making: a part of the caustic alkali combines with the sebaceous acid of the tallow, forming sebat of potash or sebat of soda—saline matters to be found, with other neutral salts, in soap-makers spent lyes.

So far from lime rendering dung more soluble, it *impedes* that process, by forming with the dung *insoluble* salts, and it otherwise injures the dung by *disengaging* and throwing *into the air* the ammoniac or *volatile alkali*, that otherwise would have combined, and have formed neutral soluble salts, with the phosphoric or oxalic acids of the dung, or other vegetable or animal matters. The mixing of *hot lime with dung* has been highly disapproved of in the preceding part of this Work; and the Author must now conclude, by observing, that the application of it, for the purpose of dissolving the *imaginary oil* contained in dung, is too injudicious a practice, as well as, in a chemical point of light, too erroneous a theory, to have been permitted to pass without notice.

ADDENDA.

## ADDENDA.

It has been neglected, under the article Peat and Peat Mosses, to state, that their waters are very injurious to the health of cattle; and that such bad effects may be prevented by collecting the rain water that may fall on the roofs of the dwelling-house and offices, into tanks properly constructed, and having no communication with the soil. Should not the buildings be conveniently placed for affording to the cattle a supply of water from the tank, or should the extent of such roofs be insufficient to collect the quantity of rain water that may be required, sheds or hovels, covered with tile, should be erected in a central field, conveniently situated for securing to the cattle their daily supply. A farther benefit will ensue by the shelter and protection that such sheds or hovels will afford the cattle. This method of collecting rain water (and which is practised in many countries) may with great advantage be adopted in the upland, chalky, or gravelly soils, or in the marshes near the sea shore, where the springs either are at a great depth, or

where

where the water is brackish. In upland situations, water may be collected into tanks in great abundance, during the rainy seasons, by leading the surface water into such receptacles, without incurring the expence of sheds or hovels; but in fens, morasses, peat mosses, and marshes or flat grounds, near the sea shore, where the soil is full of vegetable and animal matter, or where the water is brackish, a supply of water can only be obtained by the assistance of the roofs above recommended.

APPEN-

# APPENDIX.

## BARBADOES.

VOL. I. BOOK III—PAGE 345.

" THE soil in the low lands is *black*, somewhat reddish in the shallow parts, on the hills of a chalky marl, and near the sea generally sandy; of this variety of soil, the *black mould is best suited* for the cultivation of the cane, and *with the aid of manure* has given as great returns of sugar in favourable seasons, as any in the West Indies, the prime lands of Saint Kitts excepted."

## ISLAND OF GRENADA.

VOL. I. BOOK III—PAGE 376.

" To the north and the east the soil is a brick mould, the same or nearly the same as that of which mention has been made in the History of Jamaica. On the west side it is *a rich black mould* on a substratum of yellow clay.

H h 2

To

To the south, the land in general is poor, and of a reddish hue; and the same extends over a considerable part of the interior country."

## ISLAND OF SAINT CHRISTOPHER.

VOL. I. BOOK III—PAGE 429.

" The interior part of the country consists of many rugged precipices and barren mountains. Of these, the loftiest is Mount Misery (evidently a *decayed volcano*) which rises 3711 feet in perpendicular height from the sea. Nature, however, has made abundant amends for the sterility of the mountains, by the fertility she has bestowed upon *the plains*. No part of the West Indies, that I have seen, possesses even the same species of soil that is found in Saint Christopher's; it is in general a *dark grey* loam, so light and porous as to be penetrable by the slightest application of the hoe, and I conceive it to be the production of *subterraneous fires*, the *black* ferruginous pumice of naturalists, finely incorporated with a pure loam of virgin mould. The under stratum is gravel, from eight to twelve inches deep. Clay is no where found except at a considerable height in the mountains. By what process of nature the soil which I have mentioned, becomes

more

more especially suited to the production of sugar than any other in the West Indies, it is neither within my province or ability to explain."

## PAGE 430.

" I am informed, however, that the planters of Saint Christopher's *are at a great expence for manure*."

## ISLAND OF NEVIS.

### VOL. I. BOOK III—PAGE 434.

" The soil is stony; the best is a loose *black mould* on a clay. In some places, the upper stratum is a stiff clay, which requires labour; but properly divided and pulverized, repays the labour bestowed upon it."

## ISLAND OF ANTIGUA.

### VOL. I. BOOK III—PAGE 446.

" This Island contains two different kinds of soil; the one *a black mould* on a substratum of clay, which is naturally rich, and when not chequed by excessive droughts, to which Antigua is particularly subject, very productive.

The

The other is a stiff clay on a substratum of marl; it is much *less fertile* than the *former*, and abounds with an inirradicable kind of grass, in such a manner, that many estates consisting of that kind of soil, which were once very profitable, are now so impoverished and over-grown with this sort of grass,* as either to be converted into pasture land, or to become entirely abandoned. Exclusive of such deserted land, and a small part of country that is altogether unimprovable, every part of the Island may be said to be under cultivation."

## ISLAND OF JAMAICA.

### VOL. II. BOOK V.—PAGE 204.

" It may be supposed, that a plant thus rank and *succulent* requires a *strong and deep* soil to bring it to perfection, and as far as my own observation has extended, I am of opinion that no land can be *too rich* for that purpose.

"The very best soil, however, that I have seen or heard of, for the production of sugar, of the finest quality and in

---

* Perhaps the tendency which this soil has to produce this kind of grass, might be corrected by lime, alkaline or neutral salts.

in the largest proportion, is the ashy loam of Saint Christopher's, of which an account has been given in the History of that Island. Next to that is the soil which, in Jamaica, is called brick-mould; not as resembling a brick
in colour, but as containing such a d ue mxture of clay
and sand as is supposed to render it well adapted for the
use of the kiln. It is a deep, warm, and mellow hazzle
earth, easily worked; and though its surface soon grows
dry after rain, the under-stratum retains a considerable
degree of moisture in the driest weather; with this advantage too, that even in the wettest weather it seldom
requires trenching. Plant-canes in this soil, (which are
those of the first growth) have been known in very fine
seasons to yield two tons and a half of sugar per acre:
after this may be rekoned the *black mould* of several va-
rieties. The best is the *deep black earth* of Barbadoes,
Antigua, and some other of the Windward Islands; but
there is a species of *this mould* in Jamaica that is but little,
if any thing inferior to it, which abounds with limestone and flint on a substratum of soapy marl. *Black
mould* on clay is *more common*; but as the mould is generally
shallow, and the clay stiff and retentive of water, this
last sort of land requires great labour, both in ploughing

and

and trenching, to render it profitable. Properly pulverized and *manured*, it becomes *very productive*, and may be said to be *inexhaustible*".

PAGE 206.

" It is remarkable, however, that the same degree of ploughing or pulverization, which is absolutely necessary to render stiff and clayey lands productive, is here not only unnecessary but *hurtful*; for though this soil is deep, it is at the same time far from being heavy, and it is naturally dry. As, therefore, *too much exposure* to the scorching influence of a *tropical sun destroys its fertility*, the system of husbandry on sugar plantations, in which this soil abounds, is to *depend chiefly* on what are called ratoon canes," *(or sprouts of the canes formerly planted)*—these *continue* to be cut, and to produce sugar for some years: as they *decay* they are replaced by *fresh* plants. By "this" method the planter, instead of stocking " or *digging up* " his ratoons, suffers the stoles to continue in the ground, " instead of stocking, or digging up his ratoons, and *holeing and planting* the land anew."

PAGE 208.

" In the cultivation of other lands (in Jamaica especially) the plough has been introduced of late years, and

in

in some few cases to great advantage; but it is not every soil or situation that will admit the use of the plough, some lands being much too stoney and others too steep. And I am sorry I have occasion to remark, that a practice commonly prevails in Jamaica, on properties where this auxiliary is used, which would *exhaust the finest land in the world.* It is that of *ploughing*, then *cross* ploughing, *round ridging*, and *harrowing* the *same* lands from *year to year*, or at least every *other year*, *without affording manure.* Accordingly it is found, that this method is *utterly destructive* of the ratoon, or second growth, and altogether *ruinous.* It is indeed astonishing, that any planter of common reading or observation, should be passive under *so pernicious a system."*

PAGE 215.

" Hitherto I have said nothing of a very important branch in the sugar-cane planting : I mean the method of manuring the lands. The *necessity* of giving even the *best soil* occasional assistance, is *universally admitted*; and the usual way of doing it in the West Indies, is now to be described.

" The manure generally used, is a compost, formed,

" First, Of the coal and vegetable ashes drawn from the fires of the boiling-house and still-house.

I i     " Secondly,

"Secondly, Feculences discharged from the still-house, mixed up with the rubbish of buildings, white lime, &c.

"Thirdly, The refuse or field trash, i. e. the decayed leaves, and the stems of the canes, so called in contra-distinction to cane trash reserved for fuel, and hereafter to be described.

"Fourthly, Dung obtained from the horse and mule stables, and from moveable pens, or small inclosures made by posts and rails, occasionally shifted upon the lands intended to be planted, and into which the cattle are turned at night.

"Fifthly, Good mould collected from gullies and other waste places, and thrown into the cattle pens."

PAGE 217.

"But the *chief dependance* of the Jamaica planter, in manuring his lands, is on the moveable pens, or occasional inclosures; not so much for the quantity of dung collected by means of those inclosures, as for the *advantage of the urine* from the cattle (the *best of all* manures) and the labour which is saved by this system. I believe, indeed, there

are

are a great many overseers who give their land *no aid* of any other kind than that of *shifting the cattle* from one pen to another, on *the spot* intended for planting, during three or four months before it is ploughed or holed."

*Note.* "This, however, is *by no means sufficient* on plantations that have been *much worn and exhausted by cultivation*; and, perhaps, there is no branch in the planting business wherein attention and systematic arrangement, as saving both time and labour, are more necessary than in collecting and preparing *large quantities of dung* from the sources and materials before described."

PAGE 218.

"The young sprouts are at the same time cleared of weeds, and the dung which is spread round them, being *covered with cane trash*, that its virtue may not be *exhaled* by the sun, is found, at the end of three or four months, to be *soaked into* and *incorporated with* the mould."

PAGE 219.

"Such is the general system of preparing and manuring the land in Jamaica. I have been told that more attention is paid to this branch of husbandry in

some

some of the Islands to windward ; but I suspect there is in all of them very *great room for improvement*, by means of *judicious tillage and artificial assistance*."

*Note.* " It should have been observed, that it is sometimes a custom, after a field of canes has been cut, to *set fire* to the trash. This is called *burning off*; and there are managers and overseers who consider it as one of the best methods of meliorating the land. I confess that I am of a *different opinion*. Perhaps, indeed, in a moist, stiff, and clayey land, it may *do no harm*; and this negative phrase is the *only merit* I can allow it. From the usual and prevalent nature of the soil best adapted for sugar, I am persuaded that nine times in ten it is a *mischievous practice*."

FINIS.